2017 Press Conference Records of Ministry of Environmental Protection, the People's Republic of China

Ministry of Environmental Protection of the
People's Republic of China

2017 Press Conference Records of Ministry of Environmental Protection, the People's Republic of China

CEPG

Springer

Ministry of Environmental Protection
of the People's Republic of China
Beijing, China

Translated by Li Haiying, Liu Qi, Li Yutong, Liu Haiyan, Liu Meng and Nie Jia

ISBN 978-981-13-7329-9 ISBN 978-981-13-7330-5 (eBook)
https://doi.org/10.1007/978-981-13-7330-5

Jointly published with China Environment Publishing Group Co., Ltd., Beijing, China
The print edition is not for sale in China Mainland. Customers from China Mainland please order the print book from: China Environment Publishing Group Co., Ltd.

Library of Congress Control Number: 2019935826

This Springer imprint is published by the registered company Springer Nature Singapore Pte Ltd.
The registered company address is: 152 Beach Road, #21-01/04 Gateway East, Singapore 189721, Singapore

Preface

Ministry of Environmental Protection (MEP) launched the mechanism of regular press conference in January 2017. Over the past one year, MEP communicated to the public the important decisions and arrangements by the Party and the government, measures taken by MEP and relevant progress, and responded to the hot-spot issues of public concern through this mechanism, which facilitated the support and understanding of the media and public for environmental protection.

MEP held a total of 12 press conferences in 2017, basically covering all its work priorities, including environmental quality monitoring and air pollution prevention and control, ecological conservation, water pollution prevention and control, law compliance inspection, environmental technological standards, soil pollution prevention and control, international environmental cooperation, environmental policies and regulations, environmental impact assessment, integrated air pollution prevention and control in the Beijing–Tianjin–Hebei region, environmental protection planning, and central environmental inspection. The officials in charge of the above work attended the press conference. This mechanism provided the public a convenient channel to environmental policies, raised the environmental awareness, and created a favorable environment for people to engage in environmental protection.

This book has three parts in chronological order.

Part I is the records of Mr. Li Ganjie, Minister and Secretary of the Leading Party Members' Group of MEP responding to questions from media at the Press Conference held by the Press Center of the 19th National Congress; Part II is the records of 12 regular press conferences of MEP in 2017; and Part III is the records of the special press conferences, such as those on stepped-up action of comprehensive air pollution prevention and control and nuclear safety. We hope that this book will give some enlightenment to the people working in environmental protection, journalists, and readers who pay great attention to and support environmental protection.

Due to the limitation of our knowledge, there might be errors or inadequacies in the book. Your suggestions and comments are highly appreciated.

Beijing, China
January 2018

The team of compilation and edit

Contents

Part I Record of the Press Conference of the 19th National Congress of the Communist Party of China (CPC)

Records of the Press Conference on "Promoting Green Development for a Beautiful China" 3

Part II Record of MEP's Regular Press Conference

Regular Press Conference (January) 27

Regular Press Conference (February) 43

Regular Press Congress (March) 63

Regular Press Conference (April) 79

Regular Press Conference (May) 97

Regular Press Conference (June) 109

Regular Press Conference (July) 127

Regular Press Conference (August) 145

Regular Press Conference (September) 167

Regular Press Conference (October) 189

Regular Press Conference (November) 221

Regular Press Conference (December) 249

Part III Special Press Conferences of MEP

Press Conference on MEP's Stepped-Up Action for Comprehensive Air Pollution Prevention and Control in the Autumn and Winter of the Beijing–Tianjin–Hebei Region and Its Periphery 277

Press Conference on the 13th Five-Year Plan for Nuclear Safety and Radioactive Pollution Prevention and Control and the 2025 Long-Term Targets 297

Part I
Record of the Press Conference of the 19th National Congress of the Communist Party of China (CPC)

Part I
Record of the Press Conference of the 19th National Congress of the Communist Party of China (CPC)

Records of the Press Conference on "Promoting Green Development for a Beautiful China"

October 23, 2017

The Press Conference at Press Center of the 19th National Congress of CPC

The Press Center of the 19th National Congress of CPC held a press conference at 15:00, October 23. Mr. Yang Weimin, Deputy Director of the Office of the Central Leading Group on Financial and Economic Affairs, and Mr. Li Ganjie, Minister and Secretary of the Leading Party Members' Group, of the Ministry of Environmental Protection were invited to brief the journalists on the latest progress in promoting green development for a beautiful China, and respond to their questions.

Ministry of Environmental Protection of the People's Republic of China, *2017 Press Conference Records of Ministry of Environmental Protection, the People's Republic of China*, https://doi.org/10.1007/978-981-13-7330-5_1

Li Ganjie, Minister and Secretary of the Leading Party Members' Group, Ministry of Environmental Protection

Guo Weimin (Moderator): Ladies and gentlemen, and dear journalists, good afternoon. Welcome you to the sixth Press Conference held by the Press Center of the 19th National Congress. We are very pleased to have Mr. Li Ganjie, Minister and Secretary of the Leading Party Members' Group, the Ministry of Environmental Protection, and Mr. Yang Weimin, Deputy Director of the Office of the Central Leading Group on Financial and Economic Affairs, at today's press conference. They will brief you on the latest progress in promoting green development for a beautiful China.

Li Ganjie: Dear journalists, good afternoon. I am very delighted to meet you. I would like to take this opportunity to express my heartfelt thanks to your long-term support and understanding for environmental protection and ecological civilization building.

I am very honored to participate in the 19th National Congress of CPC as a delegate to the Congress, and listened to the report delivered by President Xi Jinping to the 19th National Congress. Of great significance, this report fully demonstrates the common wish of the party and the Chinese people of all ethnic groups. It is also a remarkable and epoch-making masterpiece featuring the Marxism that will serve as a political declaration and program of action for our party to make a great struggle, develop a great project, pursue a great cause, and realize our great dream.

The report comprehensively reviews the ecological civilization building and ecological and environmental protection, and lays out the key arrangements for next steps. We working in environmental protection are greatly encouraged and inspired by this report, and will strongly support and fully implement it. I would like to take this opportunity to brief you on the progress made in ecological civilization building and environmental protection.

Since the 18th National Congress, the CPC Central Committee with Comrade Xi Jinping at its core has planned and done fundamental, long-sighted, and innovative work, which promoted historic, transformational, and systemic changes taking place in both awareness and protection of ecological environment. The past five years has been the best period for ecological civilization and environmental improvement with the most profound awareness, unprecedented efforts, and fastest and greatest progress, which can be detailed as follows:

First, we have never had such a deeper understanding of ecological civilization. The entire Party and the whole people have become more purposeful and active in pursuing green development, and there is clear shift from the tendency to neglect ecological and environmental protection.

Second, we have made unprecedented efforts to prevent and control pollution. War was declared on pollution with action plans on prevention and control of air, water and soil pollution. Environmental infrastructure such as waste water and waste disposal facilities were developed at an accelerated speed. Coal-fired units with an installed capacity of 570 million kilowatts have been upgraded for energy conservation and ultralow emissions. A total of over 18 million yellow-labeled vehicles (gasoline-fueled vehicles that fail to meet the national I emission standards and diesel-fueled vehicles that fail to meet the national III emission standards) and old vehicles were phased out. Comprehensive environmental improvement projects have been carried out in more than 110,000 villages, benefiting nearly 200 million villagers. Major ecological conservation and restoration project went on smoothly.

Third, systems and institutions have been established for environmental protection at a frequency never seen before. The central leading group for comprehensively deepening reform has approved over 40 specific reform plans for ecological civilization and ecological and environmental protection, greatly facilitating green development and environmental improvement.

Fourth, law enforcement supervision is more stringent than ever before. A number of important laws have been promulgated and amended, including *Environmental Protection Law, Law on Air Pollution Prevention and Control*, *Law on Water Pollution Prevention and Control*, *Law on Environmental Impact Assessment*, *Law on Environmental Protection Tax* and *Law on Nuclear Safety and so on. The Law on Soil Pollution Prevention and Control* has been submitted to the Standing Committee of the National People's Congress for deliberation. With the new *Environmental Protection Law* passed in 2014 and entering into force in 2015, some new regulations and mechanisms established in this law well promoted the business sector in law compliance.

Fifth, the environment has been improving at an unprecedented pace. In 2016, the average PM2.5 concentration in Beijing–Tianjin–Hebei Region, the Yangtze River Delta and the Pearl River Delta was brought down by over 30%

from that in 2013 when *Law on Air Pollution Prevention and Control* entered into force. The ratio of acid rain areas to the total national land dropped from a record high of 30 to 7.2% in 2016. In fact, the acid rain coverage ratio in China was very high, even reaching over 30% in the 1990s. Our efforts have brought down both the acid rain coverage and degree of pollution. The share of surface water sections under national monitoring program recorded with a quality of Grade I, II and III was up to 67.8%. Forest coverage rate increased from 16.6% in early 21st century to about 22%.

While addressing domestic environmental issues, China is also playing an active role in global environmental governance. China has ratified or joined more than 30 multilateral environmental conventions or protocols and pioneers in international cooperation on climate change, thus becoming an important player, contributor and leader in the global development of ecological civilization.

Those achievements achieved are an epitome of the historic changes taking place in the cause of the Party and the country since the 18th National Congress. They lay a solid foundation for next steps and strengthen our confidence.

Fully recognizing the achievements made in building an ecological civilization, the report puts forward a series of new ideas, new requirements, new targets and new arrangements for ecological civilization building and ecological and environmental protection.

On the front of new ideas, it establishes the idea of ensuring harmony between human and nature as a basic policy for upholding and developing socialism with Chinese characteristics for a new era, and puts forward important conclusions that building an ecological civilization is vital to sustain the Chinese nation's development, and that man and nature form a community of life.

On the front of new requirements, it points out that the principal contradiction facing Chinese society in the new era is that between unbalanced and inadequate development and the people's ever growing needs for a better life, and that the modernization that we pursue is one characterized by harmonious coexistence between man and nature. It states that in addition to creating more material and cultural wealth to meet people's ever growing needs for a better life, we need also to provide more quality ecological goods to meet people's ever growing demands for a beautiful ecological environment.

On the front of new targets, it sets the following goals: winning the war against pollution by 2020, achieving a fundamental improvement in the environment and basically attaining the goal of building a beautiful China by 2035, developing China into a great modern socialist country that is prosperous, strong, democratic, culturally advanced, harmonious and beautiful, and achieving new heights in every dimension of material, political, cultural and ethical, social, and ecological advancement by the middle of this century.

On the front of new arrangements, it proposes to promote green development, solve prominent environmental problems, intensify the protection of ecosystem, and reform the environmental regulatory and supervisory system. Those new ideas, new requirements, new targets, and new arrangements form guidelines for developing a new model of modernization with humans developing in harmony with nature and building a beautiful China.

President Xi stressed in the report that never forget why you started, and you can accomplish your mission. The original aspiration and the mission of Chinese Communists is to seek happiness for the Chinese peoplc and rejuvenation of the Chinese nation. Good ecological environment is in the interest of Chinese people and important for our nation's long-term development. We working in environmental protection should keep in mind the requirements made by President Xi and expectations from people, keep to our original aspiration and press ahead with endeavor.

We will put the guiding principles from the 19th National Congress into action, in particular Xi Jinping Thought on Socialism with Chinese Characteristics for a New Era, act on the vision of making development people-centered, stay committed to socialist ecological civilization, vigorously promote ecological civilization building and ecological and environmental protection, so as to build a beautiful China and meet people's ever growing demands for a beautiful environment. First, we will strengthen the consciousness of the need to maintain political integrity, think in big-picture terms, follow the leadership core and keep in alignment, to fulfill the political responsibility of building an ecological civilization. Second, we will develop eco-friendly growth models and ways of life, to restore nature to its original state of serenity, harmony, and beauty. Third, we will step up efforts to solve the acute environmental problems of air, water, and soil pollution for a steady improvement of environmental quality. Fourth, we will strengthen the protection and restoration of ecosystems to increase the supply of quality ecological goods. Fifth, we will deepen the reform of the system for developing an ecological civilization to improve environmental management systems. Thank you.

Guo Weimin: Next let's welcome Deputy Director Mr. Yang Weimin.

Yang Weimin: Dear journalists, good afternoon. I am very delighted to meet you at the press conference and have this opportunity to brief you on the progress made in reforming the system for developing an ecological civilization.

China has made remarkable economic progress since the reform and opening up. However, it can not be denied that ecological environment has become a weak link in national development and "a pain spot" for people's livelihood. To beef up this weak link and cure the "pain spot", we should give priorities to both ecological civilization development and the reform of the system for ecological civilization development, with more emphasis put on the latter. Since the 18 CPC National Congress, the CPC Central Committee with Comrade Xi Jinping at its core puts the reform of the system for ecological civilization development high on its agenda. 20 meetings out of 38 ones held by the Central Leading Group for Comprehensively Deepening Reform were about the reform of the system for ecological civilization development, with 48 key reform items worked out. Among the 37 tasks for the reform of the system for ecological civilization development set up at the 3rd, 4th, and 5th plenary sessions of the 18th Central Committee of the CPC, 24 have been finished, 9 partly finished and 4 are under way, with 84 reform-related documents issued.

Since a top-level design was not well established for the ecological civilization development system, the reform of the system for ecological civilization development relatively lagged behind the reforms in other areas, especially the economic system reform. In 2015, the CPC Central Committee and the State Council issued the *Reform Plan of the System for Ecological Civilization Development* (hereinafter referred as *Plan*). This *Plan* identified the overall framework and designed 8 systems for the ecological civilization development. Up until now, the reform is going on smoothly. Among the 79 tasks set up in the *Plan* for 2015 to 2017, 73 have been finished and 6 basically finished.

The system for ecological civilization development now is developed at an accelerated speed, as the reform on the system of property rights for natural resources assets is proactively pushed forward, the system for developing and protecting the territorial space increasingly strengthened, the pilots for reforming the spatial planning system in full swing, and the system for regulating total consumption and comprehensive conservation of resources constantly strengthened. Sustained efforts are made in implementing the system of payment-based resource consumption and compensating conservation and protection efforts, the reform on environmental governance system is strengthened, the development of a market system for environmental pollution control and ecological preservation sped up, and the system for evaluating officials' ecological conservation performance and for holding those responsible for ecological damage to account is basically established.

Ecological civilization building is a profound reform in China's history of development. The five years following the 18th CPC Congress marked a period during which ecological civilization system reform was pursued with the strongest measures and fastest and unprecedented progress. This demonstrates the historic developments of the cause of the Party and the country, as well as the historic transformation of ecological civilization development system.

I will give you some examples. One is about the system of property rights for natural resources assets. In the past, there was no clearly defined ownership and explicit powers and responsibilities for natural resources except for mineral resources. This system was not changed 60 years since the new China was founded and over 30 years of reform and opening up, until the CPC Central Committee with Comrade Xi Jinping at its core initiated the reform on this system, with focus on defining the property rights for natural resources and establishing a management system. It may take time before the system of property rights for natural resources assets is fully established, but historic steps have been made.

As for land development, the nation in the past went after overdevelopment, aiming at industrialization and urbanization. This went against the laws of economic development and nature. Efforts now are being made to establish the functional zoning system, under which development activities are optimized, prioritized, limited or prohibited based on the local positioning of function. For example, Beijing is categorized as a place for optimized development. President Xi proposed to relocate Beijing's noncapital functions, so as to blaze a new trail towards intensive development and explore a model of optimized development in densely populated areas. This can be seen as another historically significant restructuring of Beijing following those in Yuan, Ming, and Qing Dynasties.

Also, almost all the government departments that have the mandate related to natural resources management set up their respective protected areas. The number and area of such protected areas were huge, but without effective regulation, they existed in name only. President Xi chaired the examination and approval of the plans for national parks pilots of Sanjiangyuan (123,100 km^2), Manchurian tigers and leopards (14,600 km^2), panda (27,100 km^2) and Qilian Mountain (50,200 km^2), aiming to protect the originality and integrity of ecological systems in those areas. This is to return a total of 21,5000 km^2 of land (equivalent to 2% of territorial space) to nature, as well as to panda, Manchurian tigers and Tibetan antelopes, leaving more clean land to our future generations. This is an unprecedented undertaking in China's history of development.

There exists the problem of overlapping in spatial planning, which means that a piece of land may be designated for different purposes of utilization by different government departments. This is a problem that we want to solve over a long time. As an important part of national governance system, spatial governance system entails a unified and complete spatial planning. As required by President Xi, different development plans at each city or county should be integrated for one. Only in this way can a plan be fully implemented. President Xi chaired the examination and approval of the plans for spatial planning pilots in Hainan province and Ningxia Autonomous Region. Those pilots show that it is feasible to have only one plan for a region.

China is implementing the strictest possible farmland protection system, ensuring that its farmland stays above the red line of 1.8 billion mu.[1] However, the protection of other ecological space is relatively weak compared to that of farmland, which leads to the encroaching of forest, wetland, grassland, and ocean for construction. President Xi said that mountains, rivers, forests, farmlands, lakes, and grasslands were a community of life. Thus, we should adopt a holistic approach to conserve them. Currently, the system of regulating total consumption is implemented for natural resources, including land, water, energy, natural forest, grassland, wetland, desertificated land, ocean, and mineral.

In the past, some places pursued economic growth at the expense of environmental protection. President Xi said that the performance of an official was not judged only by local GDP growth, and that those responsible for ecological damage would be held to account. As a series of institutional arrangements are put in place, government departments at local level now are more purposeful and active in environmental protection. This is a fundamental change.

As Minister Li said, environmental governance witnessed many historic changes. Following the 18th National Congress, the 19th National Congress again sent

[1] 1 mu is equivalent to 666.667 m^2.

out the call for accelerating the reform of the system for ecological civilization development and building a Beautiful China. This shows the will and determination of the CPC Central Committee with Comrade Xi Jinping at its core in strengthening ecological civilization building. To secure a decisive victory in building a moderately prosperous society in all aspects and embark on a journey to fully build a modern socialist China, we will win the tough war against pollution, strengthen the wink link of ecological environment, and provide more quality ecological goods to meet people's ever growing demands for a Beautiful China. As the sky gets bluer, water cleaner, and mountains greener, people can clearly see stars and hear birds chirping. In such way is harmony between human and nature achieved.

I will stop here. Next, I am ready to take your questions. Thank you.

Guo Weimin: Let us move to the Q&A session.

CCTV: I have a question for two ministers about the reform of the system for developing an ecological civilization. As the fourth batch of central environmental inspection teams got stationed at local level, a total of 31 provinces, regions, and municipalities have been inspected over the past 2 years, during which over 10,000 officials were held accountable for local environmental damage. What are your comments on this large-scale inspection? Some people also expressed their concern that pollution would come back again after inspection teams left, or that some enterprises might stop high-polluting production when being inspected, and started production again as inspection teams left. What measures will be taken to address such concern? How to make such inspection mechanism a long-term and powerful deterrence? Are there any specific measures for the next step? Thank you.

Li Ganjie: I will respond to your question. Central environmental inspection is an important move proposed by President Xi in reforming the system for developing an ecological civilization since the 18th CPC National Congress. Inspection was piloted in Hebei province at the end of 2015, and has covered 31 provinces, regions and municipalities when the 19th CPC National Congress opened. Such inspection over the past two years is very effective, highly recognized by different stakeholders. Some describe it as an undertaking to which the public give thumbs-up, recognized by the central government, supported by local governments, and effective in solving problems. I personally believe that such description is quite appropriate. The remarkable achievements made can be detailed as follows:

First, ecological and environmental protection was substantially improved and the consciousness of the need to strengthen ecological protection and pursue green development strengthened. President Xi pointed out in the report of the 19th National Congress that the entire Party and the whole country had become more purposeful and active in pursuing green development, and that there was a clear shift from the tendency to neglect ecological and environmental protection. I personally believe that central environmental inspection plays a very important role in facilitating such changes.

Second, some acute environmental problems of public concern were addressed. The inspection of the past four batches received a total of 135,000 reportings, some of which overlapped with others. We transferred 104,000 to local governments for handling after merging some overlapping ones. Up by now, 102,000 have been concluded, among which about 80,000 are related to environmental problems of waste, oil and fume, fetor, noise, enterprises which are in unauthorized locations, lack the proper certificates and fail to meet emissions standards, and black and putrid water. Thus, such inspection is well recognized and supported by the public.

Third, central environmental protection inspection promoted the transformation and upgrading of local industries. As inspection was carried out at local level, many local governments saw it as a good opportunity as well as a motivation to promote green development and pursue supply-side reform. Thus, measures were taken to strengthen pollution prevention and control and internalize environmental costs, leveling the playing field for those enterprises that comply with laws. With those enterprises which are in unauthorized locations, lack the proper certificates and fail to meet emissions standards were dealt with, the trend of "bad money drives out good" was reversed, and the industries that comply with law saw great increase in its scale and performance.

Fourth, mechanisms for environmental protection and ecological civilization at local level were improved. Inspection motivated local governments to go through the existing problems and accelerated their steps in improving relevant laws and regulations. During the process of inspection, about 300 to 400 laws and regulations were issued, which well promoted ecological civilization building and green development. I think the achievements made in the above four aspects are very prominent.

The following six features can explain why such great progress was made in central environmental inspection. First, committed to making development people-centered, we seriously took the problems of public concern and acted on solving them. Second, we strengthened the consciousness of the need to maintain political integrity, think in big-picture terms, follow the leadership core and keep in alignment, taking a clear political stand. Third, the inspection on party and government officials was strengthened to ensure that they are held equally responsible and that leading officials must take responsibility for environmental protection within their jurisdiction. Fourth, we followed a problem-oriented approach. Fifth, information was disclosed timely for the public to supervise. We made public every reporting received. Sixth, those officials responsible for environmental damage were held accountable.

The above six features are successful experience drawn from this central environmental inspection. They are in line with China's national conditions and institutional characteristics, and follow the laws of development. Thus, we will carry them on both in our future inspection and other environmental protection work.

The achievements we achieved in inspection manifest the profound and historic progress and changes taking place in the cause of the party and the country.

You asked if inspection would be continued and whether the pollution problems of some enterprises would come back again after inspection teams left. The inspection over the past two years is very effective, highly recognized by the public and bringing a lot of good practice and experience. Thus, we will definitely carry it on. After the latest batch of inspection concluded, we are now working on the following as the leading government department for inspection:

First, we will comprehensively review the past four batches of inspection to spot problems and draw experience. Based on this, we will continue to improve relevant mechanism and supporting measures for the next round of inspection.

Second, we will research and promote the formulation of laws and regulations on inspection, ensuring that inspection is carried out and sustained for a long-term by rule of law and in a standardized way.

Third, we will carry out special inspection in key regions and on key industries that targets key problems. This is being planned and may be implemented by the end of this year or early next year.

Forth, we will guide and urge local governments to establish an environmental inspection system at provincial level. This is to develop a mechanism under which municipalities and counties are inspected by provinces and provinces inspected by the central government. Now the mechanism of provinces being inspected by the central government has been basically established. Local governments are working to establish a mechanism under which the inspection and monitoring branches of environmental protection agencies below the provincial level are brought under the direct oversight of their respective provincial-level agencies, in a bid to facilitate the mechanism of municipalities and counties being inspected by province. Such mechanism under which inspection is coordinated at the national level and provincial level helps release more momentum of inspection. We welcome supervision form media. That's all. Thank you.

Yang Weimin: I fully agree with what Minister Li said. The report of the 19th National Congress pointed out that we would take tough steps to stop and punish all activities that damage the environment. This demonstrates the Party's stand and determination in environmental protection: whatever activities that damage the environment will be resolutely stopped, toughly punished, and heavily fined. I believe with a long-term mechanism established through those institutional arrangements, a social atmosphere under which none dare nor has the desire to damage the environment will be developed. Thank you.

The Wall Street Journal: Is there a contradiction between the target of improving the ecological and environmental protection, and that of securing a stable economic growth? What measures will be taken by the government of China to deal with this contradiction? We noticed that those factories that failed to meet environmental standards stopped production. Will this raise the unemployment rate? Thank you.

Li Ganjie: A journalist asked a similar question that if environmental protection would impact GDP growth at the press conference of the day before yesterday, and Mr. Zhang Yong, Vice Director of National Development and Reform Commission (NDRC)gave a very good answer. It is no denying that environmental protection will impact certain enterprises on a micro level, but in a long-term and from a macro perspective, it will promote their development instead of impacting them. There is a positive correlation between economic development, and strengthened environmental protection, green development and ecological civilization development.

President Xi reiterated in his report that lucid waters and lush mountains were invaluable assets. He also pointed out that protecting the environment was protecting productive forces, and improving the environment was developing productive forces. Such remarks were made through sound analysis and based on large amounts of successful cases. Some good cases in other countries also can be seen in China. Due to time limit, I will not go into detail.

Since the 18th National Congress, in particular over the past 2 years, environmental protection actions that we took such as environmental inspection and special campaign on environmental law enforcement in some key regions did not hinder the economic development. As you know, China's economic growth momentum is strong, which is reflected in such indicators as GDP growth, fiscal income and foreign trade.

We also kept track of economic data of the provinces where inspection was carried out. The data showed that economic indicators were in some sense better than before, not affected by inspection at all.

As for the employment, as you may know, the unemployment rate of major cities in China has dropped to the record low in recent years. This would not be possible if our environmental protection work had affected employment.

The war against pollution is both tough and time-consuming. Environmental problems cannot be solved overnight since the fundamental reasons behind them are deep-rooted. Thus, we should apply appropriate policies and approaches in this war. For example, under the "zero tolerance" policy, strict punishment is imposed on enterprises for infractions of environmental laws and regulations. At the same time, we will follow the differentiated principle and take targeted measures case by case.

Some media reported the problem of "one size fits all" (approaching each case with the same and inflexible yardstick, regardless of their respective situations). I want to clarify that we resolutely oppose such undifferentiated approach in handling polluting enterprises. Generally speaking, it only happened in a few places and has been corrected immediately once spotted. The enterprises that violate laws and regulations are allowed time for rectification, instead of being shut down immediately. Only those that heavily pollute environment and have no ability to meet environmental standards through rectification will be shut down.

It is because we follow a right track in environmental protection that our efforts are supported by different stakeholders including enterprises and local governments. As a result, environmental protection and economic growth are reinforcing and complementing each other, with a balance reached between economic growth and environmental improvement. That's all and thank you.

China Youth Daily: Deputy Director Yang mentioned that 20 out of 38 meetings held by the Central Leading Group for Comprehensively Deepening Reform were about the reform of the system for ecological civilization development. The achievements achieved in this reform are prominent. What do you think are the difficulties and problems in pursuing the reform, and what are the priorities for the next step? Thank you.

Yang Weimin: Thank you for your question. We are now exploring the right track to the reform of the system for ecological civilization development. I just mentioned that the ecological civilization system reform relatively lagged behind the economic system reform, and there were no precedents that we could learn from. For example, China is the first country to explore the functional zoning system and balance sheets for natural resource assets. We indeed encounter difficulties in pursuing this reform, and the progress in some areas is relatively slow. For example, the reform on the system of property rights for natural resources assets is now being piloted. One challenge we face is that it is very difficult to determine the ownership of water flows because the Constitution prescribes that water flows are owned by the whole people.

In addition, integration of systems and mechanisms is difficult. Government departments now have an overlap in their mandates for ecological protection and environmental governance, and this will be dealt with as the 19th National Congress identifies the overall plan for the next step. Environmental governance is regulated by corresponding laws. As you all know, the revision of a law is a complex and time-consuming process, which will slow down some reforms in environmental protection.

I think the biggest problem is that, despite the strengthened regulation by government, the role of market mechanism has not been played to its full extent. Thus, the incentive mechanism needs to be established and improved. For example, how to reduce the profits that one can get from emitting pollutants? The incentive mechanism will be the priority of the reform for the next step.

The reform of the system for ecological civilization development will be accelerated in line with the arrangements of the 19th National Congress. Following a problem-oriented approach, we will strengthen those wink links and eliminate the loopholes in environmental systems and mechanisms. Thank you.

China Meteorological News: Early this year, China Meteorological Administration and Ministry of Environmental Protection announced to jointly establish a forecast and early warning mechanism that targeted smog. What progress has been made in this regard? President Xi stressed again in the report of the 19th National Congress that we would win the battle for blue sky. What do you think is the toughest challenge for this battle? Thank you.

Li Ganjie: Thank you for your question. Forecast and early warning is the first step in air pollution prevention and control, neither of which is dispensable. A clear picture of pollutant emissions and meteorological conditions is crucial to forecast and early warning. Thus, forecast and early warning would not be possible without the close cooperation between China Meteorological Administration and Ministry of Environmental Protection.

Forecast and early warning over the past two years were getting more accurate. I think the close and fruitful cooperation between China Meteorological Administration and Ministry of Environmental Protection plays a very important role in this. What you may not know is that we established a mechanism under which experts from China Meteorological Administration and Ministry of

Environmental Protection meet regularly and mobilize their expertise to analyze pollutant emissions and meteorological conditions. Such meeting is held daily in autumn and winter. I pay great attention to daily forecast and notice that the forecast is accurate, which shows that this mechanism of experts cooperation is very effective. In addition to cooperation in forecast and early warning, we are discussing cooperation in other areas. I believe that under the concerted efforts of both sides, we will achieve more.

President Xi pointed in the report that we would continue our campaign to prevent and control air pollution and win the battle for blue sky. The wording of "win" instead of "fight" represents higher requirements for environmental protection. To explore how to win this battle, we are now studying the report to fully understand the principles and spirit behind it. At the same time, we will step up efforts in air pollution prevention and control. This is to ensure that by 2020, substantial progress will be made in pollution prevention and control, particularly in battle against air pollution. The *13th Five-Year Plan for the Protection of Ecological Environment* sets up the target that the percent of days with good air quality in 338 cities at prefecture level and above will reach over 80%, and the annual average concentration of fine particulate matter (PM2.5) in the cities that fail to meet air quality standards decrease by an average of 18% from that in 2015. We will make a good plan for achieving the target of a fundamental improvement in the environment by 2035.

We are now looking into what "fundamental improvement" means. I personally think that the annual average concentration of $PM_{2.5}$ of 35 $\mu g/m^3$ should be achieved not only at national level, but in some key areas. This is a very high standard. We will devote upmost efforts to improve air quality, in a bid to meet the people's ever growing needs for a better environment. Thank you.

Phoenix Television: I have a question for Minister Li. This year marks the last year of the Phase I for implementing the *Action Plan on Air Pollution Prevention and Control*. However, the air quality in some key cities like Beijing still fell short of the targets set up for the Phase I. Will MEP take some follow-up measures? The number of smog days did not decrease in this autumn and winter. Is it because that the shut-down policy for polluting enterprises is not stringent enough? Will more forceful measures be taken?

Li Ganjie: Based on what I have introduced about the *Action Plan*, I will offer some details to respond to your question. It has been a bit more than 4 years since the *Action Plan* was released in September, 2013, and substantial progress has been made. This shows that we have well implemented the *Action Plan* and are on the right track guided by the ideas, direction, pathways, and measures identified in the *Action Plan*.

I will give you three sets of data to show air quality improvement from 2013 to 2016. First, according to the *Action Plan*, air quality in 338 cities at the prefecture level or above are measured and evaluated based on the level of PM10 concentration. The PM10 concentration in those cities dropped by 15.5%. Second, the average PM2.5 concentration in Beijing–Tianjin–Hebei Region, the Yangtze River Delta and the Pearl River Delta decreased by 33%, 31.3% and 31.9% respectively.

Third, the percent of days with good air quality increased from 60.5 to 74.2%, an increase of 13.7 percentage points, and that of heavily polluted days decreased from 8.6 to 3%, a drop of 5.6 percentage points. Those changes reflect the prominent improvement of air quality since the *Action Plan* was put in place.

Despite the progress, we still have many problems in air pollution prevention and control, which can be summarized as follows:

First, generally speaking, air pollution is still severe, and air quality is still far from meeting people's expectations. Only 84 cities among 338 cities at the prefecture level or above (24.8%) saw their air quality reaching the standards. The national average PM10 concentration was 82 $\mu g/m^3$, 17.1% higher than the standard of 70 $\mu g/m^3$. The national average PM2.5 concentration was 47 $\mu g/m^3$ last year, 34.3% higher than the standard of 35 $\mu g/m^3$.

Second, air pollution in some key regions and during key periods become more severe. For example, North China, in particular Beijing–Tianjin–Hebei region and its surrounding areas, is frequently stroke by heavy air pollution. This got even worse in last winter, which people living in Beijing–Tianjin–Hebei region experienced. Such heavy air pollution caused great public concern.

Third, the industrial structure, energy mix, and transportation structure have been adjusted and improved in recent years, but not yet to an extent that can bring noticeable air quality improvement. Currently the share of heavy chemical industry is still too high, coal dominates the energy mix, and transport heavily relies on highways.

Fourth, lacking the consciousness of the need to comply with laws, some enterprises are emitting pollutants beyond standards.

Fifth, the pressure of air pollution prevention and control has not been fully passed to governments at local level. Governments of higher level are urgent to deal with environmental problems, while governments of local level do not recognize such urgency, thus not fully implementing environmental policies.

I believe that the problems I just mentioned are prominent in some regions. This raises people's concern of whether the targets set for the Phase I can be well achieved in the last year of Phase I and what follow-up measures will be taken. We are soberly aware of those problems and stay committed to drawing a good conclusion to Phase I in line with the requirements of the central government.

In fact, several months ago, we have looked into those problems and planned our work for the next step in some key regions, especially the Beijing–Tianjin–Hebei region and its neighboring provinces that attract great public attention. Since this area is very typical, I will brief you the relevant progress.

To deal with air pollution in the autumn and winter in Beijing–Tianjin–Hebei region and its neighboring provinces (Beijing, Tianjin, Hebei, Shandong, Henan, and Shanxi), MEP, together with 10 other government departments and the above 6 provinces and municipalities, issued the *Action Plan on Comprehensive Air Pollution Prevention and Control in the Autumn and Winter of Beijing–Tianjin–Hebei region and its neighboring provinces for 2017 to 2018* and another 6 supporting plans (1+6 documents). We have held a special press conference on this *Action Plan*.

We are working to implement this set of 1+6 documents and good progress have been made. I think the following 10 features, some of which drawn from the central environmental protection inspection, well explain the progress made.

First, priorities were highlighted. Second, a systematic and integrated approach to making and implementing policies was applied. Third, detailed plans for finishing environmental protection tasks were worked out. Fourth, governments and parties were closely inspected in fulfilling their environmental protection responsibilities. Fifth, competent staff of environmental protection was mobilized. Sixth, technologies were applied in environmental protection. We sent to every city a team of experts who would guide the local pollution prevention and control. Seventh, overall coordination was stepped up. We are working to establish an air environmental agency for Beijing–Tianjin–Hebei region and its neighboring provinces, to ensure that unified planning, standards, permits, monitoring, law enforcement and emergency response are applied. The pilots on this are under preparation. Eighth, the accountability system was quantified, which is unprecedented in China's history of environmental protection. This mechanism has taken very good effects. Ninth, information transparency and disclosure was strengthened. You can now find a lot of information about the tough battle against air pollution in Beijing–Tianjin–Hebei region and its neighboring provinces. Tenth, efforts were made to communicate environmental policies to the public and guide them in environmental protection. This is very important since it will raise people's consciousness of the importance, need and urgency of this battle, and facilitate the public participation in environmental protection. The battle against air pollution is a tough one that will go on for a long time. Thus, we should have the determination, confidence as well as patience in this battle. We should not rush for quick results, because those deep-rooted and long-standing problems cannot be solved overnight.

We are now fighting the battle against air pollution according to the overall plan. The battle started in September this year and will last to March next year. I would like to stress that this is the first phase of the battle, and we will continue this battle in the long-term. Thus, we will not stop with working out temporary measures, but explore how to establish long-term mechanisms through implementing those measures. The already established long-term mechanisms are effective in environmental protection and will be carried on. That's all. Thank you.

Guo Weimin: You just mentioned that a team of experts was sent to every city. Do you mean cities at prefecture level and above?

Li Ganjie: Yes.

Hong Kong Business Television: I have a question for Mr. Yang Weimin. Hong Kong has products in green finance certification. The direct access for mainland investors to Hong Kong bond market, dubbed as "southwards connect" of the Mainland-Hong Kong bond market connect, has not been established. Is there a timetable for this? In the following five years, will more measures be taken to promote the financial interconnectivity between mainland and Hong Kong, in particular green finance?

Yang Weimin: We still have great potential in mobilizing market mechanisms to promote green development, including green finance. The report of the 19th

National congress proposed to develop green finance. Last year, the People's Bank of China, together with other relevant government departments, has released the guidance and opinions on developing a green finance system. Relevant government departments have their own share of work in this initiative.

Preliminary achievements have been made in developing green finance. For example, China issued 134 billion yuan of labeled green bonds from January to September this year, taking a leading role with 24% share of the global total. By February this year, 21 major financial institutes saw 7.51 trillion yuan of green credit outstanding. The share of labeled green bonds at 8.8% is not high compared to China's huge scale of loan, but both the share and growth rate are growing fast. Some newly emerging innovative products, such as green index products, are expanding the channel of options for investors.

Other reforms on green finance are underway. For example, Ministry of Finance is working on the plan for establishing a national green development fund; China Regulatory Security Commission is looking into how to improve information disclosure of listed companies; and People's Bank of China is piloting the reform and innovation of green finance in some regions. International cooperation on green finance is carried out, such as the cooperation under the frameworks of China-France High Level Economic and Financial Dialogue and China-UK Economic and Financial Dialogue. Despite the achievements made, the development of green finance cannot meet the huge needs from green market. Thus, we still have great potential in this regard.

Hong Kong is one of the most important financial centers internationally. I believe that we will see more closer and mutually beneficial cooperation between mainland and Hong Kong in issuing green bonds. This will facilitate green finance development and support national green industry and projects. Thank you.

China Cooperation Times: President Xi pointed out in the report of the 19th National Congress that we will implement the rural vitalization strategy. We know that Jiangsu Province Supply and Marketing Cooperative is implementing the zero markup policy in pesticide supply and delivery, and that Heibei province is working on the utilization of excrement and waste from poultry breeding as resources, which helps build a closer link between the primary industry, secondary industry and tertiary industry. The above initiatives are all about rural development. Will new plans be issued to support the implementation of the rural vitalization strategy? Are there any new measures following the progress made over the past five years in agricultural nonpoint source pollution? Thank you.

Li Ganjie: Ecological improvement is an important target of the rural vitalization strategy. Governments at different levels put rural ecological improvement high on their agenda and have achieved some progress. As I mentioned before, the central government earmarked 37.5 billion yuan to comprehensive environmental improvement projects carried out in more than 110,000 villages since 2008, benefiting about 200 million villagers. Those projects not only helped solve many environmental problems that occurred on the people's doorsteps, such as pollution of waste water and garbage, but also promoted rural pollution prevention and control as well as ecological environment protection. Integrated into the work agenda of relevant government departments, those projects will continue, with the goal to cover 200,000 villages (about 1/3 of the total villages).

In addition, we will strengthen regulation on rural ecological environment protection, in particular on the transfer of urban pollution to rural areas. At the same time, we are working with Ministry of Agriculture and other relevant departments to prevent and control the pollution from livestock and poultry breeding, and pesticide and fertilizer, and promote the comprehensive utilization of straws.

We believe that as those efforts are put in place, the environment in rural areas will improve as the environment in urban areas, meeting the needs of people living in rural areas for a beautiful environment. That' all.

Yang Weimin: I just want to build on what Minister Li has said. The rural vitalization strategy is a highlight in the report. It's another important strategy following the strategy of building new socialist countryside to address issues relating to agriculture, rural areas, and rural people. I believe that the plan to support this strategy will be worked out at Central Rural Work Conference, and other supporting plans will also be put in place.

The rural vitalization strategy requires to build pleasant living environments. In the past, we overstressed industrial and urban pollution prevention and control, not dedicating enough efforts and funds to agricultural and rural pollution prevention and control. This has profoundly changed over the past five years.

For example, President Xi chaired the Meeting of the Central Leading Group on Financial and Economic Affairs to discuss the utilization of waste from livestock and poultry breeding as resources. If discharged into rivers and lakes without any treatment, pollutants from livestock and poultry breeding will cause severe damage to environment. Livestock and poultry breeding have been banned in some areas, such as where water sources are located. Then how to meet people's daily needs for meat while reducing pollution?

The answer is utilizing excrement and waste from livestock and poultry breeding as resources. For example, many enterprises ferment excrement of chicken and pig for biogas, and make natural gas out of biogas, which can be used in villagers' daily life and for heating in the winter. This can be seen in Beijing.

Waste sorting is also discussed at this meeting. Zhejiang province reported to President Xi the waste sorting in rural areas. Every household in Zhejiang province has two waste containers for separating biodegradable and nonbiodegradable waste. The public can easily tell the degradable from nondegradable. Waste should have been separated into more detailed elements, but this at least solves the problem of no segregation between dry waste and wet waste. Wet waste can be reused as compost.

One of important principles behind the *Reform Plan of the System for Ecological Civilization Development* is equal emphasis on urban and rural pollution prevention and control. I believe that stakeholders concerned will follow this principle to solve the environmental pollution in rural areas. Thank you.

Li Ganjie: As Deputy Director Yang said, we did not attach as much attention to rural pollution prevention and control as we did to urban and industrial pollution prevention and control. Such unbalance has become a wink link in China's ecological protection and ecological civilization development. However, environmental restoration and improvement in rural areas is very important since it determines whether villagers' needs for a better life and beautiful environment can be met and whether people have access to abundant and good-quality grain, vegetables, and water. This is about the life of people living both in urban and rural areas. Thus, we will put rural pollution prevention and control high our agenda. President Xi pointed out in the report of the 19th National Congress that we would intensify the prevention and control of agricultural pollution from nonpoint sources and take measures to improve rural living environments. This is like a starting gun that launches our work for the next step. In line with the requirements of President Xi and the 19th National Congress, MEP will work with departments concerned for dramatic progress in rural pollution prevention and control as we have achieved in other areas of ecological and environmental protection.

Jiefang Daily: In last January, President Xi chaired a colloquium on Yangtze Economic Belt. He pointed out that at the current stage and for quite a long period, the ecological restoration in Yangtze River was a task of overriding importance, and provinces and municipalities along Yangtze River should join hands for environmental protection of Yangtze River instead of excessive development. However, media reported many acute environmental problems and hidden risks along Yangtze River. Are there any follow-up plans for ecological restoration and environmental conservation in Yangtze River basin? Will more stringent and targeted measures be taken? How to balance the interest of stakeholders along Yangtze River? Thank you.

Li Ganjie: Thank you for your question. The CPC central committee and State Council have been attaching great importance to ecological and environmental protection in Yangtze River. President Xi stressed at the colloquium on Yangtze Economic Belt last January that we should give priority to ecological protection and green development in developing the Yangtze Economic Belt, and make ecological restoration of Yangtze River a task of overriding importance. He reiterated in the report of the 19th National Congress that we would facilitate the development of the Yangtze Economic Belt by promoting well-coordinated environmental conservation and avoiding excessive development.

Following this colloquium, local governments were working on environmental conservation and green development of Yangtze River and some preliminary progress has been made, such as water quality improvement. Water quality is affected by many factors, but its improvement can in some sense represent the efforts we have made. From January to September over the last year, the share of monitoring water sections recorded with a quality of Grade I, II and III in Yangtze Economic Belt improved by 4.3 percentage points. This did not come easily considering the fact that climate, hydrologic and meteorological conditions stayed unchanged. The share of water recorded with a quality of worse than Grade V dropped from 4.2% to 3.4%, a decrease of 0.8 percentage points. Those improvements show that efforts at local level are taking effect.

However, as you just said, there exist many acute environmental problems along Yangtze River, and we still have a long way to go before ecological environment of Yangtze River is restored and improved. I think problems along Yangtze River can be summarized as follows: first, urban waste water treatment facilities are far from meeting the needs of environmental protection, not to mention rural water treatment facilities; second, heavy industries are densely located along Yangtze River, which causes great environmental risks; third, nonpoint source pollution in rural areas is severe; fourth, some shoal and wetland were damaged and destroyed in recent years, which adversely impacted the service function of ecological system.

MEP fully implemented the requirement of President Xi to facilitate the development of the Yangtze Economic Belt by promoting well-coordinated environmental conservation and avoiding excessive development. The following are the items of work we have done over the past two years, and we will carry them on for the next step.

First, strengthening the top-level design for ecological and environmental protection in Yangtze Economic Belt. We worked with relevant departments to formulate and issue the *Plan on Ecological and Environmental Protection of Yangtze Economic Belt* in July. You may notice that this *Plan* makes a top-level design for ecological and environmental protection in Yangtze Economic Belt for a certain stage and for a certain period, in particular that in Yangtze River.

This *Plan* has its very unique features. It puts equal emphasis on the protection of water resources, water ecology, and water environment. Those three factors are equally important because water environment is closely related to water resources and ecology. If water resources are not developed in a sound way, and water ecology is not well protected, it is very difficult to improve water environment.

This *Plan* gives priority to coordination of environmental protection between upstream and downstream, environmental protection in key regions, in particular Panyang lake, Dongting lake and Yangtze Estuary, some key projects of ecological protection and environmental restoration, and to reform and innovation on institutions and mechanisms. Reform and innovation on institutions and mechanisms is the prerequisite for solving many environmental problems. As Deputy Director Yang said, the reform of the system for ecological civilization provides great motivation for environmental pollution control and ecological restoration. This is also the case for Yangtze Economic Belt.

In addition, the *Plan* proposes to build an environmentally harmonious, clean, healthy, beautiful, and safe river. To be more specific, developing an environmentally harmonious Yangtze River through sound and science-based development of water resources, strengthening the ecological and environmental treatment to develop a clean Yangtze River, developing a healthy Yangtze River through the restoration and protection of water ecological system, developing a beautiful Yangtze River by addressing environmental problems along banks of Yangtze River, and effectively regulating environmental risks to develop a safe Yangtze River. We believe that this *Plan* will well facilitate ecological recovery and protection of Yangtze River.

Second, central environmental inspection was carried out in regions, municipalities, and provinces along Yangtze Economic Belt. We put pollution prevention and control of Yangtze River high on our agenda and have solved some very acute problems through environmental inspection.

Third, we accelerate steps in drawing up ecological red line for Yangtze Economic Belt. Preliminary progress has been achieved and it is expected that by the end of this year, 11 provinces within Yangtze Economic Belt will see provincial-level ecological red line drawn.

Fourth, we have launched a campaign on ecological and environmental protection of drinking water sources in Yangtze Economic Belt. Under this campaign, we will treat pollution of 1320 drinking water sources above the county level in Yangtze Economic Belt over the following two years. It is planned to finish the treatment of drinking water sources in cities at prefecture level and above in Yangtze Economic Belt by the end of this year and that in cites at county level and above by the end of next year, with standards for water quality reached. This campaign is under smooth progress and we are confident in reaching the above targets.

Yangtze River is of great significant as our mother river. We will facilitate the development of Yangtze Economic Belt by promoting well-coordinated environmental conservation and avoiding excessive development, as required by President Xi, creating more favorable conditions for sustainable development of the Chinese nation.

Zhejiang Satellite TV: Zhejiang province, where the theory of "lucid waters and lush mountains are invaluable assets" originates, has always been dedicated to development of a beautiful Zhejiang. Several days ago, it was reported that Zhejiang has eliminated surface water sections under provincial monitoring program recorded with a quality of worse than Grade V. What are your comments on Zhejiang's efforts in building ecological civilization? What we can learn from it in developing a beautiful China? Thank you.

Li Ganjie: MEP and Zhejiang province held a conference on promoting national ecological civilization development about one month ago in Anji, Zhejiang. As you all know, Anji is where President Xi proposed "lucid waters and lush mountains are invaluable assets" in 2005. This conference at Anji was to communicate and promote good experience and practice of Zhejiang province in green development and ecological civilization building.

Without slackening its efforts in green development and ecological civilization building over the past several decades, Zhejiang province has seen solid and tangible progress, and accumulated many good experiences. Zhejiang province is far ahead of many provinces in water pollution prevention and control, manifested not only in elimination of surface water sections under provincial monitoring program recorded with a quality of worse than Grade V, but also that of black and putrid water. In addition, wastewater treatment facilities are widely constructed and operated in both urban and rural areas in Zhejiang. One important reason for this is that President Xi attached great importance to water pollution prevention and control when he worked in his capacity as Secretary of Zhejiang Party Committee, and this was also the case for his successors. Zhejiang has good experience in how to fully implement policies and how to pass the pressure of environmental protection to local level. Due to time limit, I will not go into detail. I advise you to pay site visits to Zhejiang and write some reports.

We have communicated and promoted those good experience at the Anji conference, and organized site visits to some regions of Zhejiang that have done good work in environmental protection. I believe that those engaged in site visits are deeply impressed by Zhejiang experience. MEP will continue to strengthen the communication and promotion of Zhejiang experience in environmental protection, not limited to water protection. I believe that as Zhejiang experience is rolled out, other provinces will achieve good progress as Zhejiang did.

China News Service: I have a question on soil pollution. Some media reported heavy metal pollution in some cities of China. Soil quality is closely related to food safety. Will MEP take new measures to tackle soil pollution and ensure food safety? Thank you.

Li Ganjie: Soil pollution prevention and control is very important because soil pollution impacts daily supply of grain, vegetables, and water. Soil pollution is harder to prevent and control than air and water pollution. This is why we need to pay more attention to this problem and make greater efforts to address it. Since the 18th National Congress, China released the *Action Plan on Air Pollution Prevention and Control* in 2013, *Action Plan on Water Pollution Prevention and Control* in 2015 and *Action Plan on Soil Pollution Prevention and Control* in 2016. Those three *Action Plans* are under implementation and I will brief you on some progress made in soil pollution prevention and control.

We saw progress in two items of groundwork established in the *Action Plan on Soil Pollution Prevention and Control*. The first is about formulation of laws and regulations. The system of laws and regulations on soil protection is not as sound as that on air and water protection, with law lacunae in this regard. The release of the *Action Plan* facilitates smooth and noticeable progress in soil pollution prevention and control. As I mentioned in the opening remarks, the *Law on Soil Pollution Prevention and Control* has been submitted to the Standing Committee of the National People's Congress for deliberation, and some other related standards and systems are under formulation or have been formulated. For example, regulations on environment management of farmland and urban polluted land have been worked out.

The second item of groundwork is the detailed survey on soil pollution. The key in this survey is having a clear picture of soil pollution conditions to make soil pollution prevention and control more targeted. MEP, together with relevant government departments, have formulated and released the *Overall Plan on National Detailed Survey on Soil Pollution*. We saw noticeable progress in the above two items of groundwork since the *Action Plan* was released.

In addition, another two work priorities have seen progress. The first is farmland management by category, which is led by Ministry of Agriculture. Ministry of Agriculture is piloting in Tianjin, Hunan, Hubei and Liaoning provinces the mechanism under which some areas are identified as where production of agricultural products is banned. At the same time, Ministry of Agriculture, Ministry of Finance and Ministry of Land Resources jointly released the *Plan on Piloting the Fallow and Crop Rotation System*. This system is being piloted in Changsha, Zhuzhou and Xiangtan of Hunan province. The second priority is management of land used for construction. Ministry of Housing and Urban-Rural Development, Ministry of Land Resources and Ministry of Environmental Protection are sharing information through national environment management system of polluted soil. At the same time, Beijing, Tianjin, Shanghai, and Chongqing have taken a lead in developing a list of pieces of polluted soil. Some places have issued a list of key enterprises subject to supervision on soil environment, in a bid to strengthen law enforcement and strictly control newly added polluted land.

Another work priority is that as stipulated by *Action Plan*, we are carrying out comprehensive pilots in 6 regions and technical pilots in 200 pieces of land over the last year. Those pilots are going on smoothly. We will work with relevant departments to strengthen target-oriented performance assessment, in a bid to make progress in soil pollution prevention and control as we have done in air and water pollution prevention and control, reducing as much as possible the adverse impact of soil pollution on the supply of grain, vegetables and water and on public wealth. We will also accelerate the restoration and treatment of polluted soil for noticeable improvement in soil environment.

Guo Weimin: Due to time limit, this press conference will conclude. This is the last press conference held by the Press Center of the 19th National Congress. I would like to thank journalists for your participation, interviews, and reports. The last group interview is scheduled in this evening and delegates to the Congress working in environmental protection will be invited. We welcome you to raise questions in this group interview. Thank you.

Part II
Record of MEP's Regular Press Conference

Regular Press Conference (January)

January 20, 2017

On the morning of January 20, the MEP held the first regular press conference of 2017 to answer the questions of the public on the hotspot topic of "environmental quality monitoring and air pollution control". The press conference was attended by Wu Jiyou, Deputy Director General of the Department of Environmental Monitoring of MEP, Liu Bingjiang, Director General of the Department of Air Environment Management of MEP, and Fu Deqian, Deputy Head of China National Environmental Monitoring Center. They released information on air quality nationwide in 2016, introduced work on environmental monitoring and air pollution control, and answered questions posed by journalists present.

Ministry of Environmental Protection of the People's Republic of China,
2017 Press Conference Records of Ministry of Environmental Protection, the People's Republic of China, https://doi.org/10.1007/978-981-13-7330-5_2

MEP January Press Conference (1)

MEP January Press Conference (2)

Liu Youbin (Moderator): Friends from the media, good morning! Welcome to today's regular press conference. Before it officially starts, I would like to say something first.

MEP attaches great importance to media and public opinion, which its Leading Party Members' Group specially studies and makes arrangements for. Minister Chen Jining has repeatedly given instructions on doing a good job in environmental news and public opinion, demanding that we respect the patterns of news dissemination, innovate methods and means, adapt ourselves to the new situation and requirements of news dissemination and environmental protection and foster an enabling environment where people in the whole society care about, monitor, and engage themselves in environmental protection. Vice Minister Zhai Qing personally coordinated coverage issues on many occasions, demanding that journalists be well received and the media be respected to guarantee smooth coverage and interviews.

As per arrangements of MEP's Leading Party Members' Group and instructions of Minister Chen Jining, after opening the official MicroBlog and WeChat account on November 22nd last year, MEP will host regular press conferences starting from this year to timely brief media friends about environmental protection and answer their questions.

The media community has long been attentive to and supportive of environmental protection. We hope that via the regular press conferences, we can provide you with more timely and effective services. We will take your opinions and suggestions with an open mind, continuously improve our work and strive to improve the quality and effect of the regular press conferences. Thank you.

In today's press conference, we will brief you about air quality as well as environmental monitoring and air pollution control nationwide in 2016 and answer your questions. We are joined by Wu Jiyou, Deputy Director General of the Department of Environmental Monitoring of MEP, Liu Bingjiang, Director General of the Department of Air Environment Management of MEP, and Fu Deqian, Deputy Head of China National Environmental Monitoring Center.

Now let's give the floor to Deputy Director General Wu Jiyou. He will introduce the air quality situation and environmental monitoring work nationwide in 2016.

Wu Jiyou, Deputy Director General of the Department of Environmental Monitoring of MEP

Wu Jiyou: Media friends, good morning. I am glad to have the opportunity to discuss air quality with you. I want to thank you for your long-term attention to and support for environmental protection and environmental monitoring in particular. Let me first brief you about the ambient air quality in 2016 and the progress in monitoring.

According to the results of 1,436 monitoring sites in 338 cities at and above the prefecture level nationwide, the overall ambient air quality improved in 2016, and the concentration of major pollutants in key regions went down year on year; however, air pollution was still grave in some regions, especially some northern ones in winter. In 2016, of the 338 cities at and above the prefecture level covered by the survey, 84 met the air quality target, accounting for 24.9%. The percentage of fine days was 78.8%, up by 2.1 percentage points year on year. The percentage of days with heavy pollution weather was 2.6%, down by 0.6 percentage point year on year; the concentration of $PM_{2.5}$ was 47 μg/m^3, down by 6% year on year, and the concentration of PM_{10} was 5.7% lower year on year. Among the first batch of 74 cities that implemented the new air quality standards, the ten best performers are Haikou, Zhoushan, Huizhou, Xiamen, Fuzhou, Shenzhen, Lishui, Zhuhai, Kunming, and Taizhou. The ten worst performers are Hengshui, Shijiazhuang, Baoding, Xingtai, Handan, Tangshan, Zhengzhou, Xi'an, Jinan, and Taiyuan.

Now I would like to shift to the three key regions. In the 13 prefecture-and-above level cities in the Beijing–Tianjin–Hebei region, the average percentage of days with air quality up to the standard was 56.8%, the average annual concentration of $PM_{2.5}$ 71 μg/m^3, down by 7.8% year on year, and the average annual concentration of PM_{10} 119 μg/m^3, down by 9.8% year on year. In the 25 prefecture-and-above level cities in the Yangtze River Delta region, the average percentage of days with air quality up to the standard was 76.1%, the average annual concentration of $PM_{2.5}$ 46 μg/m^3, down by 13.2% year on year, and the average annual concentration of PM_{10} 75 μg/m^3, down by 9.6% year on year. In the nine prefecture-and-above level cities in the Pearl River Delta region, the average percentage of days with air quality up to the standard was 89.5%, the average annual concentration of $PM_{2.5}$ 32 μg/m^3, down by 5.9% year on year, and the average annual concentration of PM_{10} 49 μg/m^3, down by 7.5% year on year.

Overall, the situation of national air quality in 2016 is positive. First, the number of cities meeting the target and that of fine days steadily increased. In 2016, out of the 338 cities surveyed at and above the prefecture level, 84 met the standards, increasing by 11 year on year; the percentage of fine days was 78.8%, increasing by 2.1 percentage points. Second, the concentration of particulate matters and the number of days with heavy pollution in the city continued to decline. The average concentration of $PM_{2.5}$ in the 338 cities surveyed was 47 μg/m^3, down by 6% year on year; the concentration of PM_{10} was 82 μg/m^3, down by 5.7%; and the percentage of days with heavy pollution was 2.6%, down by 0.6 percentage point. Third, the concentration of pollutants in key areas continued to decline. The average $PM_{2.5}$ concentration in the Beijing–Tianjin–Hebei region was 71 μg/m^3, down by 7.8% from the previous year and 33% from 2013, which was a big margin. The average concentration of $PM_{2.5}$ in the Yangtze River Delta region was 46 μg/m^3, down by 13.2% from a year earlier and 31.3% from 2013. The average concentration of $PM_{2.5}$ in the Pearl River Delta region was 32 μg/m^3, down by 5.9% from a year earlier and 31.9% from 2013.

It should be pointed out, however, that pollution in winter in the north remains severe. Analysis of monitoring data reveals that the air quality is relatively good from March to October, and days of heavy pollution are mostly in winter, especially during the heating period in the north. During the heating period from November 15th to December 31st, 2016, the average concentration of $PM_{2.5}$ in the Beijing–Tianjin–Hebei region was 135 μg/m^3, 2.4 times that of non-heating period. December alone saw five large-scale outbreaks of heavy pollution. The atmospheric environmental quality in the Beijing–Tianjin–Hebei region and its periphery, despite improvement from the previous year, is still the worst in China. In the region, the percentage of fine days in Henan Province, Beijing Municipality, Hebei Province, and Shandong Province was less than 60%. The region is home to nine of the ten cities with the worst air quality among the 74 cities covered. Moreover, the concentrations of the six criteria pollutants in the region are all higher than the national average.

I would also like to give a brief account of the monitoring work over the past year. In 2016, MEP earnestly carried out the air pollution prevention and control action plan. Centering its efforts around environmental quality improvement, it has painstakingly promoted the implementation of the ecological environmental monitoring network construction program and made progress in the monitoring of ambient air quality.

First, the power for national ambient air quality monitoring was all taken from the local level to the central level. According to the overall arrangements made for reforming environmental monitoring system and taking monitoring power to the central level, the task was completed in all of the 1,436 national monitoring stations in 2016. This has laid a solid foundation for defining the respective financial powers and expenditure responsibilities of central and local government, effectively avoiding administrative interference and ensuring the stability of monitoring data quality.

Second, a particulate matter components and photochemical monitoring network has been established for the Beijing–Tianjin–Hebei region and its periphery. Altogether 19 component stations, 23 manual sampling stations, and 31 laser radar monitoring stations have been set up to accurately determine the causes of heavy pollution weather and objectively assess the effectiveness of response measures. These stations enable the Beijing–Tianjin–Hebei region and its periphery to deal with heavy pollution weather in a more targeted manner.

Third, air quality forecast and prewarning has been vigorously promoted. An air quality and heavy pollution weather prewarning and forecast system has been established at the national, regional, provincial, and municipal level. The system has withstood the test and played an active role in improving air quality by providing key technical support for studying and judging the process of heavy pollution in winter and for its prevention, control, and emergency response.

Next, we will continue to strengthen the top-level design of environmental monitoring, vigorously promote the reform of the environmental monitoring system, organize all environmental monitoring work and further enhance our basic capabilities and performance in environmental monitoring, providing more support and safeguard for the building of ecological civilization and the work of environmental protection. Thank you.

Liu Youbin: Next, Direct General Liu Bingjiang will brief us about the implementation of the Air Pollution Prevention and Control Action Plan (APPCAP).

Liu Bingjiang: Good morning, everyone. Now I would like to talk about the progress of APPCAP. Since it was released by the State Council, progress has been made on all dimensions, and the air quality improvement has also been corroborated by data. The Chinese Academy of Engineering has organized more than a dozen academicians and more than 20 experts specializing in air to conduct an evaluation, the core conclusion from which is that the air pollution control work is going in the correct direction with powerful measures that begin to show their effects. To sum up, the following key tasks have been going relatively smooth.

First, the ministry has ensured that responsibilities are duly fulfilled. The ministry established an assessment and approval system centering on air quality improvement and implemented policies of "party and government officials being held equally responsible and leading officials taking responsibility for environmental protection within their jurisdiction". Prewarnings were timely given to provinces and municipalities with slow progress in air quality improvement. Last year, MEP had admonitory talks with the main personnel in charge from seven municipalities over their slow progress.

Liu Bingjiang, Director General of the Department of Air Environment Management of MEP

Second, all major projects are going well. The General Office of the State Council issued *Action Plan on Permitting System for Pollutant Discharge Control.* Nationwide coal-fired plants with an accumulative capacity of 425 million kilowatts have been renovated to meet ultralow emission standards, accounting for 47% of coal-fired plants. The power industry has worked very hard to achieve this accomplishment that draws the attention of the whole world. Small-scale "scattered" coal burning[1] is a major source of heavy pollution weather. In the Beijing-Tianjin-Hebei region in 2016, 800,000 households used other fuels to replace altogether two million tons of coal. Also, 3.9 million old vehicles and yellow-labeled vehicles (gasoline-fueled vehicles that fail to meet the national I emission standards and diesel-fueled vehicles that fail to meet the national III emission standards) were phased out, meeting the target set for the year. *The Notice on Implementing the National V Emission Standards for Motor Vehicles* was released and the National V Emission Standards took effect on January 1st, 2017 nationwide. Also, the National VI Emission Standards for Light-Duty Vehicles was released, keeping up with international standards.

Third, illegal environmental practices have been punished severely. Since the new *Air Pollution Prevention and Control Law* took effect on January 1st last year, all localities have markedly stepped up environmental regulation, supervising the enforcement of the air pollution prevention and control law in winter and meting out penalties by days for enterprises behaving illegally. Environmental and public security departments have carried out joint law enforcement for 17,000 times, dealing with 6,500 prominent issues and putting under administrative detention over 1,500 people involved in law-violating cases of air pollution.

Fourth, environmental administration policies have been improved. All 31 provinces (autonomous regions and municipalities) in China have adjusted the standards for collecting pollution charges, and some provinces have started to impose their own standard, which is 1.3 to 10 times the minimum charge. The central government has earmarked altogether 42.3 billion yuan as special fund for air pollution prevention and control to support key areas such as the Beijing–Tianjin–Hebei region and its periphery, the Yangtze River Delta and the Pearl River Delta. The ministry issued policies and measures such as subsidies for coal-fired plants meeting ultralow emission criteria, off-peak production of cement plants in northern China during the heating period and emission control areas for boats.

Fifth, the ministry vigorously deals with heavy pollution weather. As you can see, we meet and discuss regularly to more precisely predict the severity, scope, and duration of heavy pollution weather. By deploying high-tech means such as big data, remote sensing, and laser radar, we have enhanced our air quality forecasting ability. The various heavy pollution weather prewarning classification criteria applied in the Beijing–Tianjin–Hebei region have been unified and MEP is now the sole issuer of prewarning alerts, to which several dozen cities would initiate large-scale and high-level joint emergency response to alleviate the influence of

---

[1]Raw coal burned for heating or cooking in households, restaurants and small industrial boilers.

heavy pollution. MEP has sent inspection teams on multiple occasions to make public and seriously handle localities and enterprises that did not put in place prewarning measures or respond effectively and discharged waste illegally. Simulations run by entities such as Tsinghua University, Chinese Research Academy of Environmental Sciences and Beijing University of Technology show that timely declaring red and orange alerts curbed the fast increase of pollutants to a certain degree, as evidenced by the 10% to 25% reduction of $PM_{2.5}$ concentration in multiple cities. Thank you.

Liu Youbin: The floor is open now.

The Beijing News: Recently the public think that huge pressure from performance ranking results in fake local monitoring data. Could you tell us about the measures that MEP has implemented to ensure data integrity? Also, that a cross-regional environmental agency will be established in the Beijing–Tianjin–Hebei region has been discussed for a long time. Are there a plan and timetable for it?

Wu Jiyou: The integrity of monitoring data is the lifeline of environmental monitoring. Data integrity has a huge bearing on monitoring development and is hence always valued by MEP. The following measures have been taken. First, rolling out regulations and mechanisms. This is top-level design for prevention and control. MEP also issued *Environmental Monitoring Quality Management Plan for the 13th Five-Year Plan Period* and work plan on air quality management. Second, reforming institutions. At the end of November, 2016, the power of ambient air quality monitoring was all taken by the central level from the local level, which means the transition from "monitoring by the evaluatee" to "monitoring by the evaluator", an institutional safeguard that spares data monitoring from administrative intervention. Third, improving the management system. MEP issued *Implementation Rules on the Operation and Management of City Stations of the National Ambient Air Quality Monitoring Network* (for Trial Implementation). As the operation and maintenance entities have changed after the power was taken by the central level, we made specific requirements on data quality management by operation and maintenance companies. Fourth, improving the quality control system. We set about building an environmental monitoring quality control system with three tiers, namely, the national, regional and operation and maintenance entity tier. We adopt multiple measures to ensure data integrity. Fifth, sharpening judicial deterrence. Owing to our active coordination with the Supreme People's Court and the Supreme People's Procuratorate, *Interpretation Concerning Some Issues on the Application of the Law to Handling Criminal Cases of Environmental Pollution*, a document jointly issued by them and implemented since January of 2017, already penalizes environmental monitoring data fraud as the crime of "destroying computer information", thus effectively addressing the difficulty of convicting fraud offenders. Sixth, vigorously investigating and penalizing data fraud. We deploy multiple means including unannounced inspection and investigate and penalize each data fraud case we spot. We shall seriously handle cases of suspected data fraud and mete out criminal punishment in cases of execrable circumstances and solid evidence, creating powerful deterrence. All these measures prove effective in ensuring data integrity and strengthening monitoring quality management. Thank you.

Liu Bingjiang: After the central authorities requested the idea of the Beijing–Tianjin–Hebei region cross-regional institution be explored, MEP immediately took the lead in organizing and carrying out intense research, which is now in the stage of feasibility study. Next, MEP will speed up its work and deliver results as soon as possible.

Southern Metropolis Daily: I have three questions. First, word has it that MEP and China Meteorological Administration will jointly release heavy air pollution prewarnings. So could you please tell us more about the release mechanism? Also, will the criteria for prewarnings be adjusted? Second, I learned from the earlier introduction of air quality change that the average $PM_{2.5}$ concentration of Guangzhou dropped by 32.1% over that of 2013 and that the drop in the whole Pearl River Delta Region is quite significant as well. So what measures has Guangzhou taken to get this result? Third, many enterprises suspend production since November as required by their localities to prevent and control the more severe air pollution in winter. Since they already do so, would it be meaningful to issue red alerts requiring their suspension of production?

Wu Jiyou: Regarding your first question, due to differences between China Meteorological Administration and MEP in evaluation indicator system, classification criteria and publishing process, information released by the two may sometimes be inconsistent, causing difficulties to the public's understanding of these matters and responses by local government. Thus, MEP worked with the Chinese Academy of Sciences to develop a numerical simulation system based on air quality monitoring data and lists of pollution sources, which factors in changes in pollution sources, circumstances of diffusion, transfer, and sedimentation and particulate component monitoring and radar monitoring data to evaluate and calibrate our forecast in real time. Results of forecasting consultation will be discussed with localities. In cases of heavy pollution, we require localities to take targeted prewarning measures to lessen severity.

Currently we are discussing with China Meteorological Administration to introduce a better mechanism to make the prewarning, classification and release of heavy pollution more effective so that the public and local government can better respond.

Liu Bingjiang: The case of Guangzhou should be considered in the context of the Pearl River Delta. Officials in governments of the region, including the Guangzhou municipal government, took air pollution seriously in the early stage. They proposed the concept of atmospheric combined pollution in 2000 and accordingly made early and well-executed plans. There are mainly three reasons for the region's better performance as you described. First, industrial structure adjustment has been put in place. Now the Pearl River Delta has no steel enterprises or big heavy industry enterprises. Since 2000 the proportion of heavy and chemical industry, instead of rising, has been declining, indicative of the well-executed structural adjustment. Thanks to the policy of "vacating the cage to change bird", replacing old industries with modern ones, industries in the Pearl River Delta have gradually upgraded, basically following the same path traveled by developed countries. Second, the Pearl River Delta region planned for and well implemented energy adjustment quite early. It started to consume less coal earlier than anywhere else and its consumption of natural gas has been the highest in China since 2005. Plus, the region has effectively controlled motor vehicles. Third, the leaders in the region value environment and manage it well. Environmental laws and management policies are well-executed. Many policies conducive to environmental protection first took shape in the region. It is a good example to others.

Regarding the third question, currently localities have launched stronger measures to deal with heavy pollution weathers and protect people's health. From where we stand, the primary concern is that pollutant reduction measures initiated by red alerts must be implemented, a request that has a legal basis. The effect of reduction efforts is not measured by the number of production-suspending enterprises, but the reduction of pollutants discharged. Some localities claimed that they had suspended the production of several thousand enterprises during the red alert period, which is an intimidatingly large number. However, judging from power dispatch during the red alert periods in these localities, the power consumption reduction was usually no more than 20%. They argued that several thousand enterprises suspended production, but this is a claim that is not corroborated by the actual pollution reduction. We need to examine the actual pollution records of enterprises and adopt further management and control measures against enterprises causing severe pollution and discharging a large amount of pollutants.

On top of that, we hope localities can respond more scientifically. Winter is not the peak season for many industries such as cement and pharmaceuticals. Enterprises in these industries can totally schedule their production scientifically by reserving repair and maintenance for winter rather than summer and producing in seasons with better air quality and higher environmental carrying capacity. This is an example of scientific response. For the sake of people's health and reducing pollutants' peak, enterprises should schedule their production, repair, and maintenance reasonably and scientifically and make necessary adjustments whenever they can.

In conclusion, first, what matters is the actual reduction of pollutants emission, not the number of enterprises suspending production. Second, relevant enterprises should schedule their production scientifically to achieve precise and fine-tuned emission reduction. Also, measures implemented should be verifiable and quantifiable.

Ifeng.com: Among the key regions, Beijing–Tianjin–Hebei reduced its $PM_{2.5}$ and PM_{10} more than Pearl River Delta. However, nine of the ten worst-performing cities last year were in the Beijing–Tianjin–Hebei region. When various cities are ranked according to their reduction of pollutants, will you also factor in the economic development of their region or simply consider air quality?

Liu Bingjiang: *The Air Pollution Prevention and Control Law* explicitly requested ranking cities and naming the best ten and worst ten performers. Your question makes sense because industrial structure, energy mix, and development stages of a region have a huge influence on pollution. The Beijing–Tianjin–Hebei region is under one air flow field and its ambient air quality has always been poor with most of the 74 cities in the region being low-ranking. We also rank cities internally by the degree of pollution reduction. Cities reducing pollution by the largest margin are basically all from Beijing–Tianjin–Hebei region, after all, their absolute base is large. We select both good and bad performers. If you are among the worst ten performers, the ranking gives you pressure. If you are among the best performers, the ranking is an encouragement. It is the absolute status, not the relative, that ultimately determines how one feels about a region's pollution. But we consider both, the absolute and the relative status. Local governments are aware of this.

Wu Jiyou: I would like to add something. Our air quality ranking started in 2013. Now MEP is working on public ranking by the degree of change. We have formulated *Plan for Ranking Cities by Degree of Change in Air Quality*. Therefore, this year we will rank cities not only by their current air quality situation but also by the change year on year. The latter will fully reflect localities' progress in air pollution prevention and control.

China National Radio: It is generally believed that there are many difficulties in the implementation of the *Air Pollution Prevention and Control Law*. This is particularly true for the past several years when in winter the improvement secured in the previous months was offset. In light of this, do we have stronger measures to improve air quality in winter?

Liu Bingjiang: The assessment of the *Air Pollution Prevention and Control Law* has two targets, one being the percentage of change and the other the absolute volume of pollutants. For Beijing, it has achieved the former, so now the focus is on the latter. Its average $PM_{2.5}$ concentration in 2016 was 73 μg/m^3, down by 8 μg/m^3 from 2015, a 9.9% change. In the process of reducing the concentration from 73 to about 60 μg/m^3, you can imagine that it indeed faces some difficulties, the biggest of which is winter heating. Last year we and the local government introduced enhanced measures for 2016 to 2017, and the progress has been very good in the development of a coal free zone in the south of Beijing. The enhanced measures were introduced only last year, thus they were not powerful enough. This year we will step up our efforts. This means that to address the winter heating problem of 2017, we would adopt more extensive, targeted, and powerful measures.

First and foremost, suspension in production. Suspension of production in winter was previously successfully implemented in industries of cement, casting and bricks and tiles. In this winter, the pollution increase from domestic heating, which is an inelastic demand, should be offset by decrease in industrial enterprises as much as possible. Therefore, more enterprises will be required to suspend production to a larger extent. Second, clean heating, a request of the central government. Overall plan has been made and departments involved are coordinating with each other. Tasks are assigned to multiple cities on the transmission channel of air pollution. This year the replacement of polluting coal by clean coal will go very far for the purpose of reducing pollutants from winter heating as much as possible. Third, management and control of high-emission vehicles. We all know that diesel-powered vehicles cause much pollution. The Beijing-Tianjin-Hebei region consumes the most raw materials and auxiliary materials. For example, producing one ton of steel needs five tons of auxiliary materials. Therefore, we should focus on managing and controlling diesel-powered vehicles and cracking down on poor-quality fuel. Fourth, structural adjustment. The government has explicitly tasked cities on the transmission channel with structural adjustment as early as possible in such industries as steel, glass and cement. Fifth, response to heavy pollution weather. We have been evaluating the heavy pollution weather response plans of the "2+8" cities to calculate their share of contribution to pollution reduction when red alerts are issued. Currently the evaluation has basically been completed. Next, we will instruct these cities to revise their emergency response as soon as possible. We will also task cities with the same share of pollution reduction so that one will not wait for others to act first or expect others to contribute more. Cities will act together. This move will be very targeted and be a top priority. Sixth, multiple safeguard measures. I mentioned three tough challenges earlier, which are motor vehicles, coal for domestic heating and enterprises that are in unauthorized locations, lack proper certificates and fail to meet emission standards. Many large and medium sized enterprises have adopted high-end pollution control facilities. The small, poorly managed and polluting ones must be renovated on a large scale via environmentally friendly means to avoid the "bad money drives out good" phenomenon.

National Business Daily: Currently there are "2+18" cities that are identified as pollution transmission channels. It is said that another eight cities will be added to make it "2+26". Is it true? If so, why is it necessary to add the eight cities?

Liu Bingjiang: The "2+18" cities have been announced, so we all know about them. Judging from the reality this year, we can tell that every time heavy pollution occurs, it is not confined to the "2+18" cities. Other cities are affected as well. But which are they? Should they be included? Will the Beijing–Tianjin–Hebei region be affected? These are questions that we are still thinking about. Once pollution occurs, all the cities affected will face the same problem. If the disease is the same, the prescription will be the same. Overall, we need to consider the following factors. First, whether the potential transmission channel cities are in the same air flow field; second, whether their pollution is of a similar severity when heavy pollution occurs; third, whether their industrial structure is at the same level of development. As for the number of transmission channel cities and which cities will be added, we will make an official statement once the matters are finalized.

China Youth Daily: First, how is heavy pollution forecast? Second, once the monitoring power is taken by the central government, how will the over one thousand national monitoring stations be operated? Third, besides the three key regions, will cities with poor air quality like Xi'an and Chengdu be incorporated into the planning?

Fu Deqian: Regarding your first question, since *Air Pollution Prevention and Control Action Plan* was issued, MEP and the Chinese Academy of Sciences have established a numerical prediction system based on high-performance computing cluster, which is among the most advanced in the world. Coupled with the information from national air quality real-time monitoring network and domestic and overseas mainstream weather forecasting information, the system makes a daily analysis and pre-judgement of air quality development in Beijing–Tianjin–Hebei region and its periphery over the next seven to ten days. A mechanism of forecast and consultation of regional heavy air pollution has been put in place, which pays close attention to the likelihood, dynamics, and developments of large-scale heavy pollution in the region and predicts the heavy pollution process features and indicators. The emphasis is placed on critical processes, namely the onset, evolution, and retreat of heavy pollution, as well as key forecasts of influence scope, severity, duration, cities influenced and the peak of $PM_{2.5}$ concentration measured by the hour in cities and onset time point.

Regarding your second question, how well the monitoring stations are operated bears on the objectivity and authoritativeness of data, therefore MEP will step up its regulation and assessment of operation companies to ensure the stations are well-operated. To that end, the first measure is to strictly enforce operation contracts. Once city monitoring stations are taken over, MEP will, in strict accordance with the requirements specified in contracts, carry out regular and irregular inspection of personnel, vehicles, supporting equipment, and quality control labs in operation and maintenance companies. MEP will have admonitory talks with those that do not meet the requirements and order them to make corrections within the specified period. Regarding operation and maintenance personnel specifically, we require that they receive training and be certified before taking up their post. The number of operation and maintenance personnel and stations is also specified. Equipment-wise, all instruments and equipment entering the national stations must be strictly verified and inspected. Only qualified ones can be put into operation. The second measure is to standardize operation and maintenance. MEP has put forward explicit requirements on the operation, maintenance and quality control work of the staff and required that the companies strictly follow the relevant norms, formulate the operation and maintenance work flow and management system and strengthen the training and education on the technical expertise and work ethics of their staff to effectively enhance operation and maintenance. The third measure is to establish a supervision and assessment mechanism. MEP will combine remote monitoring and on-site spot check, regular and irregular inspection and funding support and operation assessment to better supervise and inspect the performance of operation and maintenance companies and ensure their productive work. Thank you.

Liu Bingjiang: Regarding your third question, indeed the focus of air pollution control is the Beijing–Tianjin–Hebei region and its periphery, the Yangtze River Delta and Pearl River Delta. But that does not mean other regions are neglected. Actually, MEP pays much attention to other regions as well. The Guanzhong Plain region is indeed plagued by severe pollution, a problem already revealed last winter. The south of Shanxi province also faces prominent pollution problems. To help these regions advance their pollution control progress, an important task on our shoulder is to push cities whose air quality does not meet the target to formulate planning for that, which is a responsibility of local people's governments.

Speaking of causes of air pollution in various regions, as things stand, they are pretty much similar in all regions. Polluted regions face common problems, such as an industrial structure dominated by heavy and chemical industries, an energy mix by coal and transportation vehicles by diesel-powered ones, and insufficient management and law enforcement. But if you analyze the specific causes of pollution, you will find out that each city has its own major problems. Air pollution results from multiple factors. We pay attention to cities whose air quality does not meet the target and hope to pin down enterprises responsible for it. But that has to be a scientific process. If shutting down these enterprises does improve air quality, the government would surely make up its mind to do it. This is why we require local governments to formulate plans for meeting air quality targets. We need to know their major problems to take targeted action. Many expert teams from the central government are instructing local governments to complete this task.

Xinhua News Agency: I have two questions. First, some people question Beijing's practice of placing a lot of monitoring stations in parks, wondering if they can reflect the actual conditions. So how are locations of monitoring stations determined? Second, I was with MEP during its local inspection and I found out that small-scale "scattered" coal burning was indeed a grave concern in rural areas. How much does it contribute to heavy pollution? What has MEP done for making coal clean and promoting its application? Is MEP considering the next step?

Wu Jiyou: Let me answer your first question. China started air quality monitoring in the 1970s and 1980s. After more than three decades, we have developed a sound mechanism and procedures for designing air quality monitoring stations. Technology-wise, based on our practices over the years and drawing upon the standards and norms in developed countries, in 2013 MEP released a technical specification for the layout of such monitoring stations, which made specific requirements on the location and number of monitoring stations. Management-wise, we enforce strict approval procedures for the monitoring stations. Installing, changing and adjusting the monitoring stations must conform with the specification and undergo strict scrutiny and approval.

Theoretically speaking, the more monitoring stations a city has, the denser they are, the more objectively will they reflect its air quality situation. In reality, however, we also need to factor in the size of the city and its population, the representativeness of space and economic and technical needs etc. Take Beijing as an example. Currently, it has 35 air quality monitoring stations, 12 of which are state-controlled. These stations are located in different districts in conformity with the national technical specification. During the 13th Five-Year Plan Period, we will further optimize the 1,436 state-controlled monitoring stations, possibly expanding them to make them more representative and reflect the actual air quality more truthfully. In addition, we have established 16 air background value stations and 95 regional air stations. We will also build stations monitoring components of some particular matters and photochemical situation. The addition of super stations will optimize the overall distribution of monitoring stations and enable them to better play their roles.

Liu Bingjiang: As for your second question, during the heating period the contribution of coal to heavy pollution in the whole region is the biggest. There is not one ounce of doubt about that. However, cities in the region differ in motorized vehicle parc and industrial structure, thus coal may not always be the largest contributor to heavy pollution in each of the cities. To quantify coal's contribution in the whole region is difficult as the figures of the cities are different and can hardly be reconciled. However, generally speaking coal's contribution is undoubtedly the largest by region.

If we consider each coal utilizer's case separately, we can see that pollution emission has been basically under control in power plants but it is not the same case for the small-scale "scattered" coal burning. There are not any measures to control coal for heating in winter. This is a big problem. Here is a figure: the Beijing–Tianjin–Hebei region consumes 40 million tons of coal for domestic heating every year but the figure only covers villages in city, urban–rural continuum and the vast rural areas. In other words, the 40 million tons of coal is only used for residents' heating at home but not for large-scale husbandry and livestock breeding or vegetable greenhouses.

In 2016, two million tons of coal was replaced with electricity or natural gas in the Beijing–Tianjin–Hebei region. There is a long way to go for clean energy to replace coal. This year the small-scale "scattered" coal burning that will be replaced by electricity according to plan doubles that of last year, which is a big increase. The replacement of coal by both natural gas and electricity entails transformation of grid infrastructure and peak shaving measures for natural gas. This is a boost of rural infrastructure building to a certain extent and thus needs more financial and policy support, which has been incorporated into planning. Regions with mature infrastructure such as grid and natural gas pipeline networks should take the lead and replace coal as much as possible. This is the task for this year. Another task is to strengthen the development of energy infrastructure in other regions and replace coal used for household heating and cooking with clean energy as much as possible. Besides natural gas and electricity, another effective substitute for coal is clean coal. Effective approaches of promoting clean coal have taken shape.

Economic Daily: Industrial structure adjustment in the Beijing–Tianjin–Hebei region will be a relatively long process. In light of this, what arrangements and measures does MEP have in store for controlling air pollution with advanced environmental protection technologies?

Liu Bingjiang: Adjustment of industrial structure and energy mix in the Beijing–Tianjin–Hebei region will indeed be a long process. But currently vigorously promoting and implementing environmental protection technologies can effectively reduce pollution emission. In the region many enterprises still use low-efficient pollution-treating facilities and their pollution emission remains severe. These facilities indeed should be phased out and replaced by high-end ones. For example, many experts point out that pollution control in non-power industries is backward, which in the Beijing–Tianjin–Hebei region mainly refer to coking, iron and steel, building materials, nonferrous metals, kilns, and boilers. Standards in these industries have been put in place and I think it is time to start a new round of upgrading their air pollution control technologies. Some enterprises in unauthorized locations lacked the proper certificates and failed to meet emission standards that have either no facilities or only backward ones should indeed be phased out. If they could afford to upgrade their environmental protection facilities, they could continue their business in the region; if not, they have to be phased out. Among enterprises of the same type, some have the best facilities in the country and some have backward ones. To some extent, the "bad money drives out the good" phenomenon is a salient issue. We are now implementing an all-encompassing program to get industrial enterprises to meet emission targets. Equipping enterprises, big and small, with the technologies that are basically universally recognized is a means of regulation for environmental protection.

China News Service: The materials released mentioned that only 84 cities met the air quality target in 2016. The $PM_{2.5}$ concentration in the Yangtze River Delta and Pearl River Delta has been quite low, why only 84 cities met the target?

Wu Jiyou: Not one single city in the Beijing–Tianjin–Hebei region met the target. Only several cities in the Yangtze River Delta met the target but more did in the Pearl River Delta region. Of the 338 cities monitored, most the cities that met the target are in Tibet, Yunan and Hainan and other places where air quality is good. Why the number of days when the air quality target was met was not high in the three key regions, namely, the Beijing–Tianjin–Hebei, Yangtze and Pearl River Delta region? Part of the reason is ozone, one of the six pollutants we measure. The ozone problem starts to surface in some cities now, particularly between May and October. The number of summer days when ozone concentration does not meet the target is increasing. The ozone problem is our focus in the coming stage. We are taking action to address it and will adopt measures to alleviate ozone pollution so that both the number of days and cities with air quality that meets the target will rise.

Liu Youbin: The Spring Festival is coming. We want to thank you for your great support to environmental protection over the past year and we wish you a happy new year and all the best for your family!

The press conference has come to an end. Thank you for your time.

Regular Press Conference (February)

February 20, 2017

MEP held a regular press conference on the morning of February 20, introducing China's natural and ecological conservation, and major progress in priority areas like ecological conservation red lines, nature reserves, and biodiversity conservation. Mr. Cheng Lifeng, Director General of the Department of Nature and Ecology Conservation of MEP, and Mr. Gao Jixi, Head of Nanjing Institute of Environmental Sciences of MEP, attended the press conference, introducing relevant developments and responding to the questions from the media.

Ministry of Environmental Protection of the People's Republic of China,
2017 Press Conference Records of Ministry of Environmental Protection, the People's Republic of China, https://doi.org/10.1007/978-981-13-7330-5_3

MEP February Press Conference (1)

MEP February Press Conference (2)

Liu Youbin (moderator): Good morning, friends from the press, welcome to the today's press conference, the 2nd of its kind this year.

Natural and ecological conservation is an integral part of our efforts to develop ecological civilization and protect the environment, and is also one that is closely followed by the media and the public. At today's regular press conference, we will introduce relevant developments on natural and ecological conservation and then take your questions.

Here with us are Mr. Cheng Lifeng, Director General of the Department of Nature and Ecology Conservation of MEP, and Mr. Gao Jixi, Head of Nanjing Institute of Environmental Sciences of MEP. Now I'll give the floor to Director General Cheng.

Cheng Lifeng: Good morning, my friends in the news circle!

First of all, I'd like to express my gratitude for your attention to and support for our work on natural and ecological conservation over the years. I'm very glad to inform you of the developments in natural and ecological conservation and would take this opportunity to respond to some issues of your concern.

For the past few years, the MEP has made earnest efforts to implement the decisions made by the CPC Central Committee and the State Council, achieving major headways in the natural and ecological conservation, which are specified as follows.

First, we have set and strictly enforced the red lines for ecological conservation with all-around efforts. On November 1, last year, the 29th session of the Central Leading Group for Comprehensively Deepening Reforms deliberated and adopted *Opinions on Delineating and Strictly Enforcing Ecological Conservation Red Lines* (hereinafter referred to as the *Opinions*). On February 7, this year, the General Offices of the CPC Central Committee and the State Council released the full text of this document. As a new step towards sustainable development taken by the CPC Central Committee and the State Council in a new era, the document has profound and far-reaching implications. It equipped us with new means of ecological conservation and regulations and opened up new ground for our work, which actually pointed the way of ecological conservation in China for the new era. The *Opinions* specifies the guidelines, overarching goals, and basic principles for setting and enforcing the ecological conservation red lines, pointing out the red lines are also the baseline and lifeline of national ecological security, and that the core is to regulate key ecological space with this simple "red line" to see no degradation in functions of ecosystems, no reduction in total areas of ecological space, and no conversion of the protected land into other use.

Mr. Cheng Lifeng, Director General of the Department of Nature and Ecology Conservation of MEP

The Leading Party Group of the MEP specially convened a meeting to learn this document and formulate plans for better implementation, when Minister Chen Jining stressed that we shall implement the document in the spirit of driving nails, which requires great persistence to deliver on the set goals. He also called for efforts to map out a work plan as soon as possible, and move fast towards it. Upon the release of *the Opinions*, Minister Chen published a signed article on it, and there were also widespread media coverage and interpretations of the document. I have to say, these reports and interpretations were not only brilliant but also to the point and down-to-earth. One of the most impressive reports was entitled "ecological conservation red lines should be imprinted in mind and drawn on the ground", which was keenly to the point. All these have been collected and edited by our department, and expected to be distributed in our future trainings.

Second, we have been upgrading the integrated management over the nature reserves on a continuous basis. Now there are 2,740 nature reserves covering 14.8% of the territory, including 446 national nature reserves with a total area of 970,000 km^2, with as many as 109 of them established as far back as the 12th Five-Year Plan period. A network of nature reserves that includes almost all types has balanced layout and is well-functioned has been established. Regulation over nature reserves has been constantly enhanced, with an integrated remote sensing system that connects both the space and ground information networks set in place to monitor national nature reserves regularly. Whenever suspected acts or violations are found, tough measures will be taken. We also had admonitory talks with the local governments and the authorities charged with those national nature reserves with prominent problems. The MEP assessed performance of 396 national nature reserves in cooperation with relevant State departments. There were also surveys over the basic conditions of nature reserves, and double efforts to collect, analyze and integrate the basic data, with which an IT-based platform for nature reserves management was able to set in place.

Third, we have made positive strides in biodiversity conservation. The CPC Central Committee and the State Council attached high importance to biodiversity conservation. A National Committee on Biodiversity Conservation (NCBC) was set up, with Premier Li Keqiang and then by then Vice Premier Zhang Gaoli serving as the chairperson successively. In 2010, the State Council approved and released the *China National Biodiversity Conservation Strategy and Action Plan* (hereinafter referred to as the *NBSAP*), specifying the guidelines, basic principles, strategic goals and tasks for biodiversity conservation, and setting out 35 priority areas of biodiversity conservation. The MEP has made earnest efforts to implement the *NBSAP* in accordance with the *Implementation Plan for Major Biodiversity Conservation Projects* adopted by the State Council. To be specific, we carried out biodiversity surveys and observation to lay groundwork for biodiversity conservation. We have set the boundaries of priority areas of biodiversity conservation, and made clear the key areas subject to such conservation, and also specified regulation responsibilities. We have worked with the Chinese Academy of Sciences to issue the *Biodiversity Red List of China*, making clear the key species to be closely followed and protected. In September 2016, China became a signatory country to the *Nagoya Protocol on Access to Genetic Resources and the Fair and Equitable Sharing of Benefits Arising from their Utilization*, and later National Committee for Biodiversity Conservation approved and released the *National Work Plan for Enhanced Management of Biological Genetic Resources* to tighten the regulation over biological genetic resources. Moreover, the MEP, together with five other State departments, circulated the *Circular on Strengthening the Utilization and Benefit Sharing of Biological Genetic Resources in Foreign Cooperation and Communications*.

In particular, it's decided at the Thirteenth meeting of the Conference of the Parties (COP13) to the *Convention on Biological Diversity (CBD)* last December that China would host the COP15 of *CBD* in 2020, when we will review the work over the past decade, and set forth goals towards 2030. I'd like to take this opportunity to express our gratitude to the Xinhua News Agency, People's Daily, and CCTV at Mexico for their timely coverage of the exciting news. We also saw extensive media coverage domestically which quoted from these sources and received a lot of attention. Our colleagues were all thrilled when watching on CCTV our colleague Mr. Bai Chengshou expressing gratitude on behalf of the Chinese Government at the COP13.

Fourth, demonstration for ecological conservation is booming, increasingly playing an exemplary role for others to follow. Localities are working earnestly to develop ecological civilization, practicing the philosophy of "Green is Gold". The MEP invited localities to exchange experience, whose good practice greatly facilitated the local efforts to improve the ecological and environmental quality and advance the green development. We held the Award Ceremony for Ecological

Civilization and the Discussion Meeting on Promoting Ecological Civilization, the first of its kind, commending 19 groups and 33 individuals at community level for their excellent performance in promoting ecological civilization. We called on efforts to set up national eco-cities and eco-counties, with 183 cities and counties meeting the qualifications and earning such title. We carried out theoretical research on ecological civilization, helping to establish an indicator system to evaluate the performance in attaining the goals of ecological civilization, which was recently released and circulated by the central government. The indicator system specified 23 indicators, with 11 of which associated with environmental protection.

Yet despite all these progress, we are also clear that the ecological conservation remains a challenging task. The ecosystems are degrading dramatically in some areas, the quality of biodiversity declines at an uncontrolled rate, and the violations against ecological environment remain. All these problems need to be addressed.

Looking forward, we will press ahead with the natural and ecological conservation during the 13th Five-Year Plan period, putting environmental quality improvement at the center of our work and ensuring well-functioned ecosystems as our major tasks. In this sense, we must set the ecological conservation read lines, put in place a strict regulating system, and pursue a more balanced layout of those nature reserves, thus ensuring the security of national ecological space. We will increase the biodiversity and set in place some major ecological conservation projects to facilitate the efforts in ecological conservation and restoration in the key areas. There will also be steps to introduce a monitoring and regulating system that connects both the space and ground information networks, and to further refine the policies, regulations, and laws so as to enhance the regulation and enforcement capacity. All these are ensuring the ecological security of the national territory, and producing more eco-friendly products for the people.

Thank you!

Liu Youbin: The floor is open to questions. Please feel free to raise your hands.

China Youth Daily: I have two questions. First, the *Opinion* highlights the Beijing–Tianjin–Hebei region and the Yangtze River Economic Belt as two priority regions for setting red lines by 2017. Why are the two regions prioritized? What preparations have been made for this purpose? Second, I just came across several figures, saying that the ecological conservation red lines have been delineated in 13 provinces and municipalities since 2000, whereas the guidance was not released until 2015. How is the timetable set in the previous initiatives different from that of the *Opinions*, anything needs to be revised to ensure alignment? With the term of "Beijing–Tianjin–Hebei region" once again brought up, will be red lines in the three different areas be delineated separately, or done together as a whole? Is it the same to Yangtze River Economic Belt? For the provinces along this belt that have already drawn their own red lines like Jiangsu, how do they coordinate with the others? Thank you.

Cheng Lifeng: Thank you for your questions. The *Opinions* specifies the following overarching objectives for delineating the ecological conservation red lines. Specifically, provinces and municipalities in the Beijing–Tianjin–Hebei region and the Yangtze Economic Belt will have set red lines by the end of 2017, with other localities to follow by the end of 2018. By 2020 the process will have been completed, with a basic framework of ecological conservation red lines in place. It's actually a very significant decision made by the central government to establish such a framework. Coordinated development of Beijing–Tianjin–Hebei region and the development of the Yangtze River Economic Belt are two of the major development initiatives of China. The decision to start this work with the two regions is one of the major steps to support such initiatives. In the meantime, the two regions are also the priority areas of ecological conservation, especially the Yangtze River Economic Belt, which was particularly noted by Secretary General Xi Jinping to refrain from large-scale development and let environment protection go first. We will get right start from the Yangtze River.

For the moment, localities are working to implement the *Opinions*. To facilitate their implementation, we already prepared a *Technical Guide for Delineating Ecological Conservation Red Lines* (hereinafter referred to as the *Technical Guide*), and set up an expert team to assist localities in deciding on the red lines. And preliminary decisions were made as to the red lines distribution and the key areas to be protected. Next we will work with the NDRC to make clear the procedures, methodologies, and technical indicators for delineating red lines. We're now consulting relevant State departments on the draft *Technical Guide* and plan to have it circulated to local provinces, autonomous regions, and municipalities by the end of June as required in the *Opinions*. We will work together with relevant State departments to jointly set up a steering group for drawing red lines, advising localities on the distribution of red lines, and pushing ahead its implementation at the local level. Also, we will coordinate and align the efforts of the neighboring provinces in delineating the red lines to uphold integrity of ecosystems at the regional level. Thank you.

Gao Jixi: I would like to take your second questions. First of all, we actually have more than 13 provinces and municipalities having their ecological conservation red lines set in the past. Right now almost every province is working on it. The *Opinions* recently released and the *Technical Guide* issued in the past by the MEP are consistent with each other in terms of the role of the ecological conservation red lines, the requirements and the methodologies of setting red lines. We don't see any apparent discrepancies in the two documents.

Yet given the ongoing efforts of the MEP and NDRC to develop a new *Guidelines on Delineating Ecological Conservation Red Lines*, there might be certain changes to the *Technical Guide*. Localities shall also make slight adjustments accordingly in setting the red lines.

Second, extra attention shall be given to red lines in the bordering areas of the two neighboring provinces, autonomous regions, and municipalities, to ensure the consistency and integrity of ecosystems. Though the work is done separately based on different jurisdictions, a holistic approach shall be adopted to align these efforts to ensure the consistency at the national level. In this sense, it's a two-step process. First, each province, autonomous region, and municipality set the red lines on its own. Second, provinces, autonomous regions, and municipalities compare notes with each other, and then coordinated on a national scale. The Beijing–

Tianjin-Hebei region and the provinces and municipalities around the Yangtze River Economic Belt are proceeding with red lines, and a national top-down design of it is also underway. As such, we have both the individual efforts of each administrative jurisdiction, and the regional and national coordination going hand in hand, to facilitate the implementation on the one hand, and ensure the integrity of ecosystems across regions and watersheds on the other hand.

China Daily: It has come to our attention that the red lines system is not only about red lines, but also covers those key areas to be protected. This is to a large extend overlapping with other ongoing systems on protected areas. How do you plan to deal with the overlapping areas under different protection systems? How to define the responsibilities of different authorities charged with this? Thank you.

Cheng Lifeng: It's also an issue that concerns many. Yes, we do see a considerable overlap in space between other land protection systems with the ecological conservation red lines system. Currently, in China, we have over 10,000 protected areas in various types like nature reserves, scenic spots, and forest parks, etc. All these together contribute to China's current land protection system and play an essential role in conserving biodiversity and upholding national ecological security. They are as well the major ones to be protected under the red lines system.

We saw some analysis of the *Opinio*ns, saying that "drawing red lines lays the groundwork, while the key is to enforce them", which in my opinion is quite to the point when it comes to how we implement the red lines system. Here I'd like to share with you some good local experience.

Hainan has developed the following good practices in setting red lines. First, there's systematic and science-based evaluation to identify the key areas in need of immediate protection prior to setting the red lines, for example, those with important ecological functions and ecologically sensitive, ensuring all essential ecological elements are taken into account in these areas.

Second, local urban spatial planning was accommodated to the set red lines. The ecological red lines were set through science-based evaluation, and based on which, development activities like urbanization, industrialization, and resource utilization are planned. In this way, the role of red lines in setting limits for development activities is truly brought into force.

Third, the red lines system was used as a powerful tool to advance other spatial planning on environmental protection so far yet to be fully implemented. The regimes of function-based zoning, ecological function zoning, and environmental protection have all set the scope and targets for ecological conservation. Now the red lines system has actually translated all these general requirements into specific protected areas.

Fourth, conflicts in separate management of land and sea were addressed. Management authority has long been unclear in the nearshore and the land–sea transition zone, leading to overlapping responsibilities. The red lines system enables a holistic management approach to these areas, effectively addressing this long-standing problem. All these suggested the importance of a science-based approach developed by Hainan province in setting the ecological conservation red lines, which are quite implementable and can be drawn upon by other localities.

You also mentioned the responsibilities of different State departments and localities, which are also specified in the *Opinions*. It says the local Party committees and local governments take primary responsibility in enforcing the ecological conservation red lines. And relevant State departments should, based on their specific responsibilities, tighten regulation over it, coordinate and guide such efforts, and conduct inspection and supervision over law enforcement on a regular basis, working together to ensure no ecological conservation red lines are crossed. It's hence made clear that it's the local party committee and government that bear the main responsibility in setting and enforcing the ecological conservation red-lines. It also articulates the needs to clearly define the roles of different departments, strengthen interdepartmental coordination, and promote collaborative efforts from both the central and local authorities. Thank you.

China Business News: Last year, the central environmental inspection teams have investigated and dealt with quite a few violations on nature reserves. What has the MEP done in upgrading the comprehensive management of nature reserves, and what's the plan for the future? Thank you.

Cheng Lifeng: Thank you for your questions. People are very concerned about the nature reserves management. It's made clear in the *Regulations on Nature Reserves* issued by the State Council that the MEP is responsible for the comprehensive management of nature reserves. With this mandate, the MEP, with the support of the competent department in charge of nature reserve-related matters, has made the following efforts to manage nature reserves in a comprehensive manner.

First, we've worked out the policies, standards, and technical specifications. We proposed to the State Council and worked with relevant State departments to develop a raft of policies and documents for strengthening nature reserves management. These include *Circular of the State Council General Office for Better Management of Nature Reserves*, *Administrative Regulations on the Redefinition of National Nature Reserves*, and *Circular on Further Strengthening the Supervision and Management of Nature Reserve-Related Development Activities*, to name just a few. So far we have basically put in place the regulating systems respectively on supervision and inspection over nature reserves, standards for establishing nature reserves, scientific investigations, and ecological and environmental inspection.

Second, we advised the State Council on approval of establishing new nature reserves and making adjustments to the existing ones and released to public the information concerning the coverage of national nature reserves and function-based zoning. The MEP has organized experts panel to evaluate and review national nature reserves for 6 times and has submitted to the State Council 23 proposals on the establishment of a total of 379 national nature reserves since the evaluation and review mechanism was introduced.

Third, we launched inspection to various nature reserves. We introduced an integrated remote sensing system that connects both the space and ground information networks to better monitor the nature reserves. As soon as problems are identified, we will act immediately and urge for rectification. Based on the 2015 monitoring results, we launched on-site investigation, identifying 6 nature reserves with prominent problems. We then held admonitory talks with their respective local municipal government, provincial department charged with relevant nature reserve issues, and their managing institutes. Also, another four national nature reserves were made public for their poor management.

Fourth, we've put in place an assessment system for better management of national nature reserves, together with the competent departments charged with land resources, water resources, agriculture, forestry and ocean, and the Chinese Academy of Sciences. As of the end of 2017, we assessed the performance of 31 provinces, autonomous regions, and municipalities in managing a total of 396 national nature reserves.

The greatest problem facing China's nature reserve management is what you just mentioned, the tensions between the protection and the development. The central government reiterated the importance of conserving natural resources and ecological environment, which in itself a red line that can never be crossed. Yet environmental violations remain, with more development activities increasingly eroding the limited ecological space and nature reserves, weakening their ecological functions and lowering the ecological values.

Lately, we saw a number of environmental violations concerning the nature reserves disclosed by the media, receiving much attention from the society. It's also a reminder to us as the department responsible for the comprehensive management of nature reserves to perform our duties well and better regulate those nature reserves. In the future, in line with the request of our minister to act promptly, identify issues quickly and resolve it soon, we will act more proactively instead of reactively, identifying the problem before it gets worse and taking immediate measures to tackle it. We have set the following priorities.

First, we will tighten regulation. We will continue with the remote sensing monitoring of nature reserves. Starting from 2017, the national nature reserves will be so monitored twice every year, one in the first half and the other in the second, and the provincial ones once a year. We will act promptly whenever problems are found. In the meantime, we're drafting the *Measures for Monitoring Human Activities in Nature Reserves with Remote Sensing and Verifying and Handling with Relevant Issues*, to make it a long-term mechanism and done regularly. In particular, we will closely follow and monitor on a long-term basis those nature reserves that are of great values and receiving a lot of attentions from the public. Once severe environmental violations are detected, we will request for rectification and hold those relevant parties accountable, by means of holding admonitory talks with relevant authorities, circulating the found problems to the public, and strengthening public supervision of local authorities in rectifying problems within the set time limit. We will open to public the government information on comprehensive regulation of nature reserves.

Second, we will improve assessment of the management performance. We will further improve the joint assessment mechanism. Apart from the regular assessment held annually, a few areas are prioritized for strengthened assessment including nature reserves in the Yangtze River Economic Belt, and Beijing–Tianjin–Hebei Region and cross-regional ones, which are also highlighted in the red lines system.

A nature reserve monitoring and assessment network will be established to track the development of the nature reserves across China.

Third, we will lay more solid groundwork. We will set up a database and a regulatory platform for comprehensive management of nature reserves. In collaboration with relevant departments, we will review and decide on the boundaries of nature reserves and the function-based zoning, as well as collect and integrate data on remote sensing imaging, ground control, and remote and real-time monitoring, so as to establish a tiered and interconnected IT-based management framework at the nature reserve, local and national level. All these will enable our work more science-based, IT-based, and detail-oriented.

Fourth, we will promote the relevant legislation. As is known to all, the *Regulations on Nature Reserves* was promulgated over two decades ago, too outdated to meet the needs now. We have another important task to do this year, which is, to evaluate the implementation of the *Regulations on Nature Reserves*. This will be done by Nanjing Institute of Environmental Sciences together with China University of Political science and Law and other partner organizations. We will put forward the recommendations on nature reserves-related legislations, particularly on the accountability for environmental violations, based on some new developments in the pilot projects on national parks. In the meantime, we will work with relevant legislative bodies to facilitate the nature reserve-related lawmaking, so as to provide legal safeguards for a sound and sustained development of nature reserves.

At last, I sincerely hope my friends from the media to continue to pay attention to and support the nature reserve management, and fully play out your role as a supervisor. With the above measures in place, plus the media and public oversight, the violators against nature reserves will definitely be forced back. As such, a firewall will be established to maintain the ecological security of nature reserves. Thank you.

Nanfang Metropolis Daily: I have two questions. The first is about nature reserves. It's pointed out in the *Outlines of the 13th Five-Year Plan for Protecting Ecological Environment*, there will be another 30 to 50 national nature reserves established by 2020 with the intent to increase its total share to 14.8%. Judging from current progress, we have basically attained this target. Will that signal changes to the existing nature reserves in the future, for example, cutting the total areas or the number of existing reserves? The second question is about pilot projects on national park system. For the moment, NDRC is taking the lead in the piloting work, with MEP taking part and providing guidance to three of these pilots, including Sanjiangyuan (source of the Yellow, Yangtze, and Mekoong rivers) pilot in Qinghai province. How is the national park different from other protected areas we already have? Are they functioning to conserve the biodiversity or else? Thank you.

Cheng Lifeng: Thank you for your questions. I'd like to update you on relevant developments and some adjustments to nature reserves. Over more than six decades' development, the total area of nature reserves has taken up 14.8% of the national territory, so far the largest nature reserve system in the world. Given

China's national conditions and the social and economic development, this share is reasonable with both its number and total coverage at a relatively high level.

Of all types of nature reserves in China, the ones at national level are the most significant. New national-level nature reserves are designated with a view to filling in the gap of inadequate protection. This will also help balance their spatial layout and better protect them. Usually, a new national-level nature reserve is established by upgrading one at the provincial level. In this sense, a new national nature reserve generally won't involve the increase of the total area of nature reserves nationwide.

Actually, the public are supportive to upgrading nature reserve from a lower level to a higher one, but a bit concerned about re-delineating them. Here I'd like to brief you on the requirements and latest progress in re-delineating nature reserves. Development of nature reserves is gradually shifting the focus from speed and scale to quality and cost-benefits. Some nature reserves established at the early stage followed the policy of "quick fix", where forceful measures were taken to protect the endangered areas. They might work quite well at a time, yet also coupled with problems like improper delineation and poor function-oriented zoning. This is particularly true in the densely populated villages and towns, where farmlands and commercial forests with less ecological values are also included into the nature reserves. This affects the local residents' work and life and undermines standard management of nature reserves. Given that, appropriate adjustments can be made, yet must be on the conditions of protection first.

Such adjustments can only take place in a few cases and follow through strict procedures. The State Council circulated the *Administrative Regulations on the Redefinition of National Nature Reserves* in 2013. It spells out the general requirements as being prudent in making adjustments, adopting category-specific measures, tracking the whole process, and strengthening accountability.

Adjustments can only be made under the conditions when, first, natural environment is changing, including encouraging the expansion of national nature reserves, as can be demonstrated in many practical cases; second, human activities are active; third, it's in conflict with major national projects. The *Regulations* also have specific provisions in respect of setting a deadline for adjustments, highlighting areas to be further protected, refining the procedures of adjustments, and strengthening the accountability.

With all these, the State Council has actually sent a very clear message that the nature reserve with the set limits and zoning can never be adjusted easily unless there's a must for it. In view of such adjustment needs, decisions shall be made carefully and the adjustments must be science-based.

The national park system is an integral part of ecological civilization institutions setup. The *Opinions on Accelerating the Development of Ecological Civilization*, and the *Integrated Reform Plan for Promoting Ecological Progress*, issued by the CPC Central Committee and the State Council, are the top design for the national park system. As you mentioned, a leading team for the pilot projects on national parks was established at the nation level, led by the NDRC, and joined by the other 12 State departments and the concerned localities. The *Plan for the Pilot Projects on National Parks* was also circulated later, highlighting 9 priority areas for such piloting including Sanjiangyuan, Shennongjia, Mount Wuyi, Qianjiangyuan, Mount Nan, the Great Wall, Potatso, Giant Panda, and Amur Tiger and Amur Leopard national parks, which are in the 12 pilot provinces and municipalities including Qinghai, Hubei, Fujian, Zhejiang, Hunan, Beijing, Yunnan, Sichuan, Shaanxi, Gansu, Jilin, and Heilongjiang. As a member of the said leading team, MEP has taken the following steps to advance the pilot projects.

First, the MEP worked together with other departments to draft the *Plan for the Pilot Projects on National Parks*, which was then circulated and implemented upon the approval of the CPC Central Committee and the State Council. Second, the MEP was fully engaged in reviewing and offering technical guidance to the implementation plans of the 9 pilot national parks. Third, the MEP, together with the State Commission Office of Public Sectors Reform, NDRC and Ministry of Land Resources (MLR) have conducted field trips to the three pilot areas of Sanjiangyuan, Shennongjia, and Qianjiangyuan, reviewing and supervising the ongoing pilot projects to ensure steady development are made at the local level. Fourth, the MEP launched relevant research on how to establish and manage national parks, and produced a preliminary research report.

So far, the pilot plans of Sanjiangyuan, Giant Panda, and Amur tiger and Amur leopard national parks have been deliberated and adopted by the Central Leading Group for Comprehensively Deepening Reforms. Other pilot plans also have been adopted and ready to put into practice. In accordance with the overall plan, MEP will take the lead in guiding and supervising the pilot projects on Sanjiangyuan, Shennongjia, and Qianjiangyuan. There will also be inspection tours in these areas before April. As specified in the plan proposed by the Leading Group, the 9 pilot projects will be concluded in the first half of the year. Based on the experience learnt from these pilots, a master plan for establishing a national park system will then be developed.

The master plan is expected to be a top design of the national park system, and will follow through what General Secretary Xi Jinping emphasized at the 12th session of the Central Leading Group on Financial and Economic Affairs that "national parks are established to preserve the authenticity and integrity of national ecosystems, and are natural heritages for generations to come, instead of profiting from their tourism. This is the most basic point that we must bear in mind". He also added, "we should protect the sites in immediate needs of protection, and better coordinate the cross-regional and interdepartmental issues generating from that". Hope the master plan will help address your concerns. Thank you.

Xinhua News Agency: I have two questions. The first one is about the adjustments to nature reserves. You've pointed out that we have strict regulations over adjustments to nature reserves, but it seems not the case for the Mount Qilian National Nature Reserve. We found that it has been re-delineated about four times, and some of the development projects inside the nature reserve were already approved before the adjustments. What's your view on these legacy issues and how do you plan to solve them? The second one is on the remote sensing monitoring results on nature reserves published by the Department of Nature and Ecology

Conservation of the MEP. It said that 33 nature reserves were heavily affected by human activities and 89 moderately affected with obvious negative affects to be seen. It also indicated that values and functions for a few nature reserves are impaired, and some ecological elements under protection saw significant reduction in number or went extinguished. Could you brief us on the developments of those nature reserves heavily affected? Thank you.

Cheng Lifeng: I already clarified the procedures and requirements for adjusting nature reserves, and some news media also reported on the Mount Qilian incident, which have been identified as the main targets of central environmental protection inspection.

I can assure you we will not tolerate any illegal activities against the central area and buffer area of a nature reserve, and they are serious violation of laws. Development activities inside the central and buffer areas of a nature reserve should be banned by all means. We will step up environmental inspection, and conduct regular inspections coupled with remote sensing monitoring, so that illegal development activities can be found at an early date and immediate measures taken to better manage the nature reserves.

As a matter of fact, the MEP as well as the other State departments and provincial governments have all set up specific enforcement and supervision systems targeting the 446 national nature reserves across China. Information including the development of nature reserves, the frequency of human activities and the problems monitored are all forwarded to the local authorities. Please be assured all these problems will be properly dealt with through strengthened supervision and inspection. For the next, the MEP while pushing forward red lines work, will also set up a regulating platform on it. It's expected to be put in trial within the year, mainly to oversee, inspect, and monitor the problems in areas under the red lines system. Immediate steps will be taken once problems are found. For legacy issues, we will make a list and urge the local governments for rectifications one by one within a prescribed period of time.

In the first half of last year, we monitored how human activities are changing in the total 446 national nature reserves between 2013 and 2015, and made public the monitoring results. In the meantime, we requested localities to verify and confirm the found problems, investigate the violations, and hold those violators accountable. Some rectification efforts are already underway. Based on the monitoring results, there are 33 nature reserves heavily impacted by human activities, with 4 suffering the most including Moschus Moschiferus National Nature Reserve in Baishan city of Jilin province, Mount Jiugong National Nature Reserve of Hubei province, Weining Grass Sea National Nature Reserve of Guizhou province, and Zhangye Heihe River Basin Wetland National Nature Reserve of Gansu province. Site investigations were then launched by the environmental protection authorities of the four provinces, together with local governments and relevant departments. They have made clear the rectification measures, institutes and persons responsible, and deadlines for rectification, which will be reported and submitted to the MEP by the end of May. We will follow up on any new developments in their work, and make public in a timely fashion.

As I just mentioned, we will introduce a system to monitor, investigate and deal with implications of human activities on national nature reserves with remote sensing technology. Last year, we released the remote sensing monitoring data between 2013 and 2015. In the near future, we will announce the data for 2015 and 2016, circulate the found problems to relevant State departments, and order the concerned localities to make rectifications.

I'd like to reaffirm here that those national nature reserves that always found with problems will be the key targets of such monitoring and regulation. There will also be post inspection on them. We will make public and hold accountable those authorities charged with the nature reserves with multiple problems and failing to take effective measures in accordance with relevant regulations of the central government. Thank you.

Economic Daily: As you just mentioned, it's announced last December that China would host the COP15 of the *CBD*. Could you brief us on developments of the biodiversity conservation in China? And what do you expect from COP15?

Cheng Lifeng: The two questions you raised are indeed very important, with the first touching upon domestic efforts and the second on international convention implementation, which are closed connected. China is among the countries with richest biodiversity in the world. Since the 18th National Congress of the CPC, our leaders have made important instructions on multiple occasions on enhancing biodiversity conservation. Never in history has biodiversity conservation been given such greater attention, neither such stronger steps be taken. Under the leadership of National Committee for Biodiversity Conservation, the local areas and State departments have picked up the pace to implement the *NBSAP*, and implemented key projects on biodiversity conservation, with notable progress made.

To date, the forest coverage in China has been raised to 21.66%, the grassland coverage 54%. The terrestrial protected areas spread over 1.7 million km^2, about 18% of the territory, meeting the goal of 17% by 2020 specified by the *CBD* ahead of schedule. Over 90% types of the terrestrial natural ecosystems, 89% types of the wildlife communities were included in national key protection programs, and most of the key natural relics have been protected under the nature reserves system. The populations of Giant Panda, Amur Tiger, Crested Ibis, Tibetan Antelope, and other rare and endangered species that live in the wild all have seen steady growth.

China's achievements in biodiversity conservation have been widely acknowledged by the international community. My fellow colleagues told me that at the COP13 of *CBD*, when the chairman announced "the hosting country of the COP15 goes to China", rounds of applauds echoed at the venue, and the heads of several delegations walked to our delegation, shaking hands to extend congratulations. The members of the Chinese delegation all felt a deep sense of pride and honor. China's success in winning the host of COP15 should be attributed to, first and foremost, the growing comprehensive strength of China, and then our achievements in biodiversity conservation and the contributions to biodiversity conservation in the world.

Now I'd like to share some information on the conference of parties (COP) to the *CBD*. *CBD* is the world's most important convention on biodiversity conservation. It was effective as of Dec. 29, 1993 and has by far a total of 196 parties. COP 15 is the first time for China to be the host since it acceded to the *CBD* 25 years ago. As the paramount meeting under the framework of the *CBD*, the COP is held every two years. The COP15 will be a milestone in the history of the Convention and it means a lot for China as the host country. It will wrap up the global biodiversity conservation work in previous decade and set forth the objectives and strategies for the biodiversity conservation around the world by 2030. By then six to eight thousands delegates representing the *CBD* parties, the NGOs and the overseas news press will come to the COP15, which will become a major diplomatic event held in China in the arena of ecological conservation and environmental protection.

China will have completed the building of a well-off society by the year 2020, also a time when the COP 15 would be hosted. It thus becomes a good timing for China to contribute China wisdom and China solution to global biodiversity conservation, and also a platform to tell China stories, and seek a pathway to promote ecological progress all over the world in collaboration with other countries. Now we are three years away from the COP15, leaving with us a very tight schedule actually. Especially considering it involves a spectrum of sectors and issues at both national and international levels, the workload is no doubt massive. As practitioners of ecological and environmental protection, we're deeply honored and encouraged to be part of the preparatory work, and more importantly, feel a heavy responsibility on our shoulders. We already kicked off the preparation work, and will work with relevant State departments and other international and national stakeholders to develop strategic objectives and action plans for the global biodiversity conservation for the next decade.

I'd also like to take this opportunity to invite my friends at the news circle to contribute your wisdoms and help with the promotion of our preparatory work. Thank you.

Ifeng.com: I'd like to go back to the ecological conservation red lines. As you said earlier, the local Party committees and governments should assume the responsibility for enforcing the ecological conservation red lines in areas within their respective administrative jurisdiction. Are there any systems in place to ensure red lines are not crossed? Some localities remain follow the traditional mindset of developing economy first, which undermines the red lines system in one way or another. Are there any measures in place to prevent this, such as off-office auditing, accountability system, or linking with one's promotion? Do we have any detailed plan? What are the supporting measures that the MEP is working on at the moment? Thank you.

Gao Jixi: As important as it is to draw the ecological conservation red lines, it is even more important to exercise rigorous management over them. The *Opinions* already puts forward specific measures to guide our efforts. Some local areas also developed their own local management measures in delineating the red lines, which laid out effective measures to keep the red lines.

As to how to ensure effective efforts at local level, we will have monitoring and assessment on a daily basis as soon as the red lines are all decided. By introducing the integrated remote sensing system that connects both the space and ground information networks, we are able to track the red lines development, and then assess the performance in accordance with what is stated in the *Opinions* as "to see no degradation in function of ecosystems, no reduction in total areas of ecological space, and no conversion of the protected land into other use". The monitoring and assessment results will be the major indicator to assess local officials' performance, and those with poor performance will be held accountable. In summary, for the management of ecological conservation red lines, we highlight both the step to set the red lines and the efforts to enforce it to ensure they are not be crossed, which calls on efforts from both the public and the local governments.

Chinese Political Consultative News: My question is for Mr. Gao. Will any adjustments be allowed once the red lines are delineated? If yes, how are the adjustments made? Thank you.

Gao Jixi: The *Opinions* makes it very clear that the ecological conservation red lines are the bottom lines for upholding national ecological security, thus must be made mandatory and left no room for further bargaining. If adjusted at one's will, they wouldn't be called red lines. They are also known as lifeline, a term also used to regulating the use of farmland, thus won't subject to any arbitrary changes given the paramount significance of protecting one's lifeline. Meanwhile, the *Opinions* also explicitly stated that the ecological conservation red lines, if involving people's well-being and major infrastructure construction, can be re-delineated. Yet such adjustments must be based on scientific assessment, and follow through strict rules. It's then made obvious that the ecological conservation red lines, once finalized, can hardly be re-delineated in principle. This message is clear that the ecological conservation red lines are the bottom lines, thus is binding, and also a lifeline, and thus won't be easily changed.

Cheng Lifeng: I would add that there are very strict restrictions on re-delineating red lines. If there were, then it must have followed the requirements that red lines can only be expanded rather than removed. Given the nature of red lines as the bottom line for safeguarding national ecological security and the lifeline, once it's removed, it will definitely exceed the safety limits and threaten the life. In the meantime, with a growing awareness of ecological conservation, the central government will channel more inputs in and more capable of conserving the ecological environment, and more efforts will be made to restore the ecosystems. All these will increase the total protected areas, and will all be protected under the red lines system to improve the integrity of the ecosystems. As a practitioner of the natural and ecological conservation, we surely hope more lucid waters and lush mountains can be preserved for generations to come.

The Paper: MEP is the department for comprehensive management of all types of nature reserves, and also the competent department for nature reserves such as Jiangsu Yancheng Wetland National Nature Reserve for rare birds. In this case, MEP is both a player and a referee. How do you deal with this dilemma and who will oversee the performance of environmental protection authorities? Thank you.

Cheng Lifeng: You did raise a very good question. Since the 1980s, relevant authorities including environmental ones at all levels have set up a number of nature reserves where immediate protection was needed to save those endangered ecosystems, and gradually put in place a standard management and development system. Such effort filled in the gap and helped to further refine the protection system. The Art. 8 of the *Regulations on Nature Reserves* stipulates that protecting nature reserves require both comprehensive management from the competent department and sector-specific management from concerning departments. Also, Para. 4 of this Article provides that the local people's government at or above the county level shall be responsible for the setting up institutions for managing nature reserves and defining their responsibilities, and final decisions will be made by the provincial people's governments based on different local circumstances.

The Yancheng Wetland National Nature Reserve you mentioned is comanaged by both Jiangsu provincial environmental protection bureau and Yancheng municipal government, with the latter taking the primary responsibility for daily management and regulation. I'd like to clarify here that MEP is not the competent department of any nature reserves.

In recent years, we also ordered some nature reserves with local environmental protection bureau or agency as their competent authority, such as Xilin Gol Nature Reserve in Inner Mongolia and Mashan Nature Reserve in Shandong, for rectification within a set time limit and strengthened public oversight over this process. Yancheng Wetland National Nature Reserve we just mentioned was also enlisted as one of major targets of central environmental protection inspection last year, and also ordered for rectification.

From now on, we will continue to step up regulation over the nature reserves. Whenever illegal activities are found, investigations and punishments will be instituted right away. I'd like to take this opportunity to invite the media and the public to help oversee the performance of nature reserves management at all levels. Thank you.

Caixin Media: As you mentioned earlier, the only adjustment to ecological conservation red lines is to expand it instead of removing it. Mr. Gao, could you share with us how the ecological conservation red lines are delineated and done in a science-based approach, are there any guidance on it, how it relates to local economic development, and how eco-compensation is implemented, so as to better guide the local efforts?

Gao Jixi: About how to delineate the red lines in a scientific approach, earlier MEP developed the *Technical Guidance for the Delineation of Ecological Conservation Red Lines*. It specified how to delineate red lines for localities to follow. The first thing to delineate red lines is to make clear the following three questions, namely, what the red lines intend for, where we put them, and how we put them.

We have put the following types of protection areas under the red lines system, first, the areas with important ecosystem functions, so that sound eco-products are provided; second, the areas with very important biodiversity, so that natural resources and species are preserved; and third, the ecologically and environmentally sensitive and vulnerable areas, with the purpose of mitigating the natural disasters and safeguarding the ecological security.

Taking the Hunshandake Desert in Inner Mongolia as an example, its ecological conservation value is not as important as other protected areas, yet can be the source of sandstorm plaguing the entire north China once damaged. Given its considerable environmental impact, it's necessary to include it in the red lines system. As for how to implement it, we usually assess and grade the protected areas based on its ecological importance, and include those important areas inside the red lines. It's rather technical, and we can provide some paper materials if anyone is interested.

About the relation between ecological conservation red lines and economic development, we heard a lot saying that the red lines may affect local economic development, which, in my opinion, is misleading. For a very long period of time, environmental protection had been generally considered against economic development. I definitely don't think so. Environmental protection, ecological conservation, and ecological conservation red lines never go against economic development.

There are two issues involved. One is the relation between local interests and overall interests. Taking the Yangtze River headwaters as an example, grazing in the headwaters may produce benefits for local areas, but damages the environment which offsets the economic interests. We heard complaints that ecological conservation red lines hindered the development, yet this is only true at the very local level, which can be resolved through ecological compensation, as a way to balance the local interests with the overall interests. For the red lines, we eye for the overall interests.

The other is how to evaluate it. When it comes to the negative effects on economic development, it only saw economic returns from developing the protected areas, yet left out the ecological value, which would be much higher if those areas are protected. For example, we've evaluated the economic value and the ecological value of per mu land of the Shennongjia National Nature Reserve, with the latter turning out to be higher. This actually reveals that when talking about the relation between ecological environment and economic growth, people usually fail to see the whole picture and the overall benefits.

Many are interested in the eco-compensation, particularly how much would be compensated. Jiangsu and Fujian province have developed their own parameters in the evaluation, which indicated how much would be compensated in each case such as the woodland, wetland, and grassland. In terms of where the money comes from, usually the localities will spend a considerable amount of fiscal revenue on ecological compensation. For example, Jiangsu has allocated 2 billion yuan each year. Meanwhile, there are over 40 billion yuan of transfer payment from the central government going to key areas with significant ecological functions every year. In the future, the transfer payment will increasingly go to the protected areas under the red lines system, which will minimize the negative effects on those protected areas that are restricted from large-scale development activities. Then how to make full use of such compensation? There are two major use, one is direct compensation, and the other is investing in industries suited to local circumstances so that localities can have stable sources of revenue, which will not only benefit the ecological conservation but also the local economic development.

I'd like to share with you a good example. Wuzhong county of Jiangsu province, saw an increase instead of reduction of local residents' incomes after the ecological conservation red lines were set. That's because they have developed industries that were more suited to the local resource endowments, which advanced local economic development.

Liu Youbin: This concludes today's press conference. Thank you.

Regular Press Congress (March)

March 20, 2017

On the morning of March 20, the MEP held the monthly regular press conference to brief the overall developments in prevention and control of water pollution in China and the progress in key tasks such as the ecological and environmental protection along the Yangtze Economic Belt, treatment of black and putrid waters, and prevention and control of water pollution in the agricultural sector and rural areas. Zhang Bo, Director General of the Department of Water Environment Management of MEP attended the press conference and responded to the questions. This is the first time that the press conference is live by picture and text on the MEP official MicroBlog.

Ministry of Environmental Protection of the People's Republic of China,
2017 Press Conference Records of Ministry of Environmental Protection, the People's Republic of China, https://doi.org/10.1007/978-981-13-7330-5_4

The March Press Conference (1)

The March Press Conference (2)

Liu Youbin (Moderator): Dear journalists, good morning. Welcome to the March press conference of MEP.

The State Council released the *Action Plan on Water* April, 2015, which attracted broad public attention. We invited Mr. Zhang Bo, Director General of the Department of Water Environment Management to today's press conference. He will brief you the implementation of the *Action Plan* and respond to your questions.

Zhang Bo: I would like to thank media for the attention paid to water pollution control. I will not go into detail about the *Action Plan* since materials have been distributed to you. Next, I will take your questions.

Liu Youbin: The floor is open for questions.

National Business Daily: The safety hazards threatening the drinking water sources that are distributed in the chemical industrial clusters along the major river banks cannot afford to be overlooked. The industrial pollution and chemical enterprises are mentioned a lot when it comes to the protection of the source waters. Then what measures will be taken to protect the source waters? Are there any problems or difficulties in such protection?

Zhang Bo: The protection of drinking water sources has always been the top priority of the environmental protection effort, and the local governments have attached great importance to it. The available monitoring data indicated fairly high standard attainment rates among the drinking water sources. Over 90% of drinking water sources in cities at prefecture-level and above met national water quality standards. Still, there are some hazards that are not negligible. The problems you just mentioned are partly attributed to historic reasons. For example, in some cases, the enterprises took root in some places where growing populations afterwards required water supply, which necessitated the designation of water sources nearby. Such historic problems arising from the economic and social development indeed contradict with the protection of the source waters nowadays. Furthermore, some local governments do not pay adequate attention to such protection. As a result, there are still some companies and sewage outlets in breach of laws inside the source water protected areas. The crux of the protection of drinking water sources hinges on standardized development and law compliance inspection. First of all, we should delineate source water protected areas and strictly implement the protection systems, by no means allowing new polluters especially heavy polluters to sit inside the protected areas. Otherwise, we will swing our "sword" without any hesitation.

Second, we should take a stepwise approach to straighten out and remove the existing hazards and request local Party committees and governments to properly handle the historic problems with economic, social, and environmental factors taken into consideration. Local governments should set a bottom line for source water protection. If any pollution problem is spotted in a source water protected area, they should either urge the enterprises located inside the protected area to relocate or find alternative water sources. They should not hesitate on this matter and we will do our part in supervision. Thank you!

Director General Zhang Bo, Water Environment Management Department of MEP

The Beijing News: I have a question about the river chief system. Minister of Water Resources also talked about this during this year's plenary sessions of the National People's Congress (NPC) and the National Committee of the Chinese People's Political Consultative Conference (CPPCC) (Two Sessions). We saw some fast developments in this front, and as far as I know, MEP and Ministry of Water Resources jointly printed and distributed some implementation plans. What are the roles of the two ministries in introducing the river chief system? How do they play their parts in the cooperative water pollution control? And how are they held accountable for failing to do so?

Zhang Bo: The river chief system is an important work mechanism established on the Chinese context as a systematic solution to water resource, water ecology, and water environment problems. The core idea is to enable the Party committees and governments at all levels to assume the main responsibility for protecting the environment, and to be more specific, to assign a leading Party/government official the responsibility for pollution control of a certain river segment. However, the river chief system does not mean it is a single person's responsibility for the river management. It is a coordination mechanism for Party committees and governments, especially the leading officials to bring the political system into full play and mobilize all departments and the whole society to jointly engage in water pollution prevention and control.

The efforts made by MEP in the river chief system can be summarized as following. As required by the CPC Central Committee and the State Council, the Ministry works with water resources and other departments concerned to implement the river chief system, and such implementation has become one of the targets for the central environmental inspection. So far, good progress has been made in most of the local places.

The implementation of this system is integrated with the law-based pollution control, by following the *Environment Protection Law*, the *Water Law*, and the *Water Pollution Prevention and Control Law* which is being amended and establishing sound standards and specifications.

Also, such implementation is integrated with the science-based pollution prevention and control. The *13th Five-Year Plan for the Protection of Ecological Environment* has been released, and a plan for water pollution prevention and control in key watersheds is about to be released soon. Attainment plans should be mapped out for non-attainment waters, in accordance with the above two plans as well as the *Action Plan on Water*. As we build our houses brick by brick, we treat water pollution in watersheds by conducting projects one after another. Only by following scientific approaches can we move forward with the control of such pollution in watersheds.

Moreover, such implementation is integrated with deepening reforms. The Central Leading Group for Comprehensively Deepening Reforms recently adopted a plan for piloting the establishment of watershed-specific environmental regulation and enforcement institutions, in accordance with the requirements of the reform of the system for developing an ecological civilization. This plan proposed unified planning, standards, monitoring, environmental impact assessment and law compliance inspection in watershed management. The deepening reforms will provide strong support for implementing the river chief system. Thank you.

People's Daily: I have another question about the river chief system. People were not enthusiastic about the appointment of a river chief in Zhejiang province, so some counties and cities became creative and began to contract river chiefs, which made it possible for an average citizen to be a river chief, in addition to a villager chief or town mayor. What is your opinion about this? Has it occurred to you that this may happen?

Zhang Bo: The local governments have been creative in implementing the river chief system, and their experience matters. We have launched surveys at local level to spot good experience and practices that can be rolled out on a national scale. The river chief system does not mean treating river pollution singlehandedly by the river chief. A Party secretary as a river chief can mobilize more resources than deputy party or government chief. It will be the rule of man if a river can be well managed only by a Party secretary as the river chief. Instead, the river chief system is a coordination mechanism under the leadership of Party committees and governments, with the collaboration of all departments and participation by the whole society. Only in this way can we give full play to advantage of the political systems. If we rely on a single person who can mobilize enough resources in the river chief system, we deviate from the path we are supposed to take.

China Youth Daily: You just mentioned the possibility of establishing watershed-specific environmental monitoring and enforcement institutions. Can you give us more details? What will the management framework for this be like? Also, we have seen setback in pollution prevention and control of Erhai Lake. Can you take this as an example and analyze China's progress in lake pollution prevention and control?

Zhang Bo: The prevention and control of water pollution within a large region should be pursued through that of watersheds, while the prevention and control of watershed pollution through that of smaller regions within the watershed. When we monitor a section of a river, there might be a large catchment area in the upstream of the monitored section, so the control of water pollution has to be based on watershed instead of administrative region. Otherwise, the monitoring and assessment standards may vary from one administrative region to another. In my opinion, unified planning means a watershed-wide water environment protection plan should be developed for a watershed and such plans should be developed and issued by superior governments or institutions. Unified assessment standards do not mean setting an unified upper limit on pollutant discharge for the whole watershed. Instead, it means distinguishing between the environmentally sensitive and insensitive areas based on the environment carrying capacity of watersheds and establishing pollution discharge regulatory mechanisms commensurate to the water quality goals.

Unified monitoring means "who assesses the water quality monitors it". This help ensure the accuracy and reliability of data. The results of water quality monitoring are an important standard in assessing the performances of local Party committees and governments. Thus, the assessment of such performances will be undermined if relevant data is inaccurate.

Unified environmental impact assessment is not about taking back all the assessment powers, but about unifying the assessment rules and mechanisms.

About the unified compliance inspection, the environmental inspection of watersheds should observe the ecological laws and follow a systematic approach. It means more than monitoring the pollution discharge. Otherwise, compliance inspection is just a theory instead of real action. We should combine pollution control with ecological conservation in our efforts to protect the environment of a watershed.

There are some challenges in protection of Erhai Lake despite that the national and local party committees and governments have attached great importance to it. The scale of local livestock and poultry breeding and farming has expanded to accommodate the needs from inbound population who travel to visit Erhai Lake, which leads to the increase of energy consumption and pollution emissions. Therefore, local governments should have a clear picture of their respective environmental carrying capacity and how many people and industries that such capacity can support. Environmental carrying capacity is in part depends on local lifestyles. For example, carrying capacity is higher in places that follow intensive development model and lower in places with extensive development model.

I'm discreetly optimistic about pollution prevention and control in lakes across China. I believe we can treat the pollution but the effort is very demanding. We should fully understand the systematic approach behind the idea proposed by President Xi that mountains, rivers, forests, farmlands, lakes, and grasslands were a community of life. Pollution prevention and control of river basins lies on three important factors: state points of water quality, emission reduction, and capacity expansion. Concrete and practical efforts have been made in pollutant emission reduction, reflected by sound regulations and standards, regulation approaches and pollution treatment facilities, while laws and regulations on environmental carrying capacity and water ecological protection function are more like principals, standards and regulation approaches are not exercisable, and relevant responsibilities are not clearly defined, which renders water ecological environment more a concept instead of real action.

We need to highlight both the emission reduction and environmental capacity expansion for the next step. Measures will be taken to contain and prevent the following activities: shoals and wetlands are destroyed; banks of lakes and rivers are hardened out of needs of human activities; reservoirs are built excessively to accommodate water-intensive activities, cutting the access of rivers, lakes, and wetland to water replenishment even in flood season; aquaculture is developed as the water quality of rivers improved, which creates new pollution; application of dredgers to dig out spiral shell turns river bed into like desert. To deal with those problems, a priority in this amendment of *Water Pollution Prevention and Control Law* is to strengthen ecological protection of watershed.

China News Service: My question is about the surface water quality. What is the percentage of the monitored sections that meet Grade IV national standard? Can you please brief the regional imbalances in water environment protection?

Zhang Bo: *Action Plan on Water* sets out water protection objectives specific to waters of good quality and bad quality. The waters that meet Grade I, II, or III standard are considered as good and those that fail to meet Grade V are considered as bad. The percentage of water of Grade IV is changing: it decreases as the percentage of water of Grade I, II, and III increases, and increases as that of water of worse than Grade V decreases. The annual water quality goals set for 2016 by the *Action Plan* had been achieved, as 67.8% of monitored sections attained Grade I to Grade III standard, up over the previous year, and the share of waters that failed to meet Grade V standard dropped.

The progress in water environment protection varied from region to region, with setbacks in some regions. We gave guidance to such regions and supervise how the spotted problems were dealt with. Last year, we held admonitory talks with local government officials in two cities where noticeable setbacks in water quality were spotted. This was reported by CCTV and had a wide social influence. We now analyze water environment situations monthly, give alerts to those regions with setbacks, and issue notification of criticism. If the water quality in those regions do not improve or even get worse within three months, MEP will take such measures as issuing notification of criticism, holding admonitory talks, limiting the approval for new projects, and reporting to the media. I call for the local governments to analyze the water environment situations monthly to find out the problems timely, as MEP has done. Only in this way can those regions that lag behind in water protection, monitored sections with setbacks, and those enterprises that emits or discharges illegally can be spotted timely. In addition, we should mobilize the public for strengthened public supervision. I believe that the water quality targets and priority tasks established in the *Action Plan* will be finished.

CBN Daily: I still have a question about the black and putrid waters. The total amount of black and putrid waters increased rather than declined, and the treatment progress went slowly, especially in key regions. We are only halfway from the deadline set for treatment, but more than halfway from the targets. Also, when engaged in environmental protection inspection, I found out that the rivers in the outskirts of cities, for example, Shantou, were all black and putrid. So how do we plan to treat such waters?

Zhang Bo: The *Action Plan on Water* identifies the treatment of black and putrid waters as one of priorities and gives specific deadlines for phased tasks. The goal set for this year is arduous: basically eliminating black and putrid waters in some key cities by the end of this year.

This effort is being carried out by the Ministry of Housing and Urban–Rural Development with the collaboration of MEP, and the latter is making effort in the following three fronts. First, monitoring on black and putrid waters is strengthened and their treatment is incorporated into the standard attainment plans for waters. MEP monitors 1,940 sections under national monitoring programs, identifies regions that suffer from poor water quality based on monitoring data and urges them to make attainment plans, under which the treatment of black and putrid waters is listed as a key target.

Second, black and putrid waters are a priority for central environmental inspection. Black and putrid waters are caused by discharging untreated waste water directly into the environment due to lack of adequate environmental infrastructures. It is not easy to get them clean, especially in old towns and urban built areas. In those areas, the pipelines might not be established in the most optimized way considering house relocation. It is not all about money. We launched central government environmental inspection to facilitate the treatment of such waters. According to information from Ministry of Housing and Urban-Rural Development, quicker treatment progress is seen in the inspected places.

Third, environmental satellites are employed to locate black and putrid waters that have not been involved in the list of black and putrid waters for treatment. For example, we spotted 272 places of suspected black and putrid waters in 20 cities through satellite technology, among which 96 places were confirmed and put into the treatment list. At the same time, we developed an official WeChat public account to mobilize the public to engage in urban water environment management. We received a total of 3600 reportings on this platform and have dealt with 3474 ones, among which 316 were about black and putrid waters. It was verified that 50 were new additional black and putrid waters and had been put on the treatment list. The combination of satellite technology with public supervision helps spot and identify new black and putrid waters and those not found in the first place. And those waters that have been treated up to environmental standards are still subject to satellite monitoring and public supervision; if they worsen again, they will be put into the treatment list. These are the reasons why the number of black and putrid waters increased despite strengthened treatment efforts.

To summarize, we take three approaches in black and putrid water treatment: strengthened monitoring and incorporation of the treatment in attainment plans, central environmental inspection, and screening and monitoring through satellite application and public engagement. We will continue to work with relevant departments in black and putrid waters treatment. This is a demanding task that deserves great attention from the local level. The local governments have to accomplish their assignments by the end of this year; otherwise, restrictive measures will be taken.

Shanghai Securities News: I have a question about priorities of water environment management. Permits for pollutants discharge will have been granted to the printing and dyeing industry, and a national platform for information on the permits will have been established by the end of this year. Could you give more information? Why the printing and dyeing industry is the priority in granting the permits, and where are we with this?

Zhang Bo: There are ten water-related key industries that need to obtain such permit because they may lead to water pollution, for example, the paper making, printing, and dyeing industries.

We're working on the permitting system in those ten key industries, in a bid to gradually extend this system to other industries. The permitting system is the core for the management system of stationary pollution sources. Officials from MEP have written many papers on this and thus I will not go into detail. Our department is going to straighten out and improve relevant standards this year in order to make the permitting system more complementary with the watershed-wide standards and industrial standards. Only in this way can we establish a control and management mechanism of stationary pollution sources consistent with the basin-wide water quality improvement goals.

China Daily: I'd like to ask about the protection of the Yangtze Economic Belt. Minister Chen mentioned some threats to the river at this year's Two Sessions. Can you specify the pollution problems of the river and this year's work plans to address those problems?

Zhang Bo: The total phosphorus has become the top pollutant to the mainstream of the Yangtze River, which suggests some underlying issues. First, COD and ammonia nitrogen are no longer the top pollutants in the Yangtze River basin, an indication that over years of treatment, the industrial and urban pollution represented by COD and ammonia nitrogen has been curbed.

However, we cannot afford to overlook the industrial and urban pollution and still have a long way to go. It should be noted that nitrogen and phosphorus that in the past were key pollutants have become the top pollutants of Yangtze River. There are two main causes for this. The first one is about the nonpoint source pollution in agricultural sector, which contributed around 70% of the total pollution in the river. Statistics showed the application of the fertilizers amounted to 60.22 million ton in 2015, up by 45% from the year 2000.

More efforts are needed to control pollution from nonpoint agricultural sources in spite of the many measures already taken by agricultural departments.

The second cause is about water ecological conditions. Land reclamation activities have been rampant in the Yangtze River basin, especially the middle and lower reaches where the wetlands are often turned into farmlands. For example, my recent inspection tour to Chaohu Lake found the lake has shrunk to over 700 km^2 from more than 2,000 km^2 due to the loss of shoals and wetlands. Such loss has considerably undermined the environment carrying capacity of the river and reduced the local environment's self-purification capacity which was previously supported by the shoals and wetlands. We should pay close attention to address the two concerns. Thank you.

Science and Technology Daily: My question is about ecological compensations. Beijing, Zhejiang, and many other provinces already commenced the province-wide ecological compensations. Can you brief us on the progress in implementing this mechanism and what is the plan for the next step?

Zhang Bo: There're two kinds of ecological compensations. The first is the ecological compensations across administrative boundaries. For example, when the water intake is in the lower reaches of a river, we require the government in the upper reaches to take necessary measures to ensure the water quality of drinking water sources in the lower reaches. The government in the lower reaches oversees the water quality of the upper reaches and compensates appropriately. We coordinate the administrative regions concerned to sign compensation agreements, for example, the provinces in the upper reaches and lower reaches of Xin'an river, Jiuzhou river, Ting river, and Han river.

The other one is the ecological compensations within administrative boundaries, which technically are not ecological compensations. Taking Beijing for example, the city-wide water environment compensations include the monthly computed amount targeting cross-boundary sections and the annually computed amount targeting waste water treatment tasks. If a district improves the water quality at a local river section, the municipal government will reward the district; otherwise, the government will levy fines. Those districts that finish the key tasks for waste water treatment will be rewarded, while those that fail will be reported and fined. The assessment and calculation information will be made public. Such incentive and disincentive mechanism not only have an influence on the image of the government, but also reflects its internal management, thus helping mobilize the enthusiasm of the Party committees and governments at all levels to treat water pollution.

This is a long-term mechanism that can achieve great results with little effort. That's the reason why we should make greater efforts to replicate it on a national scale.

China Securities Journal: One more question about the black and putrid waters. You just mentioned that by the end of this year the municipalities directly under the central government, the provincial capital cities, and the cities separately listed in State plans will have basically eliminated such black and putrid waters. We know that its volume is very huge. Is there any special fund to support this?

Zhang Bo: The treatment of black and putrid waters is a considerably demanding task with great workload. You may find the specific figures in the materials you got, including the total number of black waters, the number of those that have been well treated, and the number of black and putrid waters for which treatment plans are developed. The money is indeed a problem, and the funds will be raised from multiple sources, such as the application of comprehensive water price (the sum of fees for waste water treatment, water resource consumption and actual costs of water treatment per ton.).

Local governments at all levels should also spend more in this effort, but it is still not enough. Local fiscal fund will be stretched if it is the only capital source for the treatment. As required by the 18th National Congress, we should give a full play to the decisive role of the market and a better play to the role of the government. Market-based mechanisms should be implemented in waste water treatment fees and water prices. At the same time, government should also earmark fund and step up regulation to treat the black and putrid waters once and for all.

RMZXB: About the control of nonpoint source pollution in the agricultural sector, we know that many rivers are polluted with excessive phosphorus, even some drinking water sources. Can you specify the initiatives for control of phosphorus pollution? In our interviews, some local government officials believe the phosphorus discharge standards are too strict and should be revised. What do you think of this?

Zhang Bo: The phosphorus pollution comes from multiple sources. We will work with the agricultural department to control nonpoint source pollution in the agricultural sector. As the competent environmental protection department, we will make greater efforts in control of point sources of such pollution, for example, the municipal waste water treatment plants and industrial enterprises. The phosphorus discharge standards include those for nitrogen and phosphorus. However, the control of nitrogen and phosphorus and relevant law enforcement are not as stringent as that of COD and ammonia and nitrogen because the latter has attracted much more attention. For the next step, we will more strictly enforce the nitrogen and phosphorus discharge standards in the said treatment plants and industries.

We may approach the nonpoint source pollution in the agricultural sector from two pathways. The first is to integrate it with the initiatives for the green development of the agricultural sector, the development of beautiful villages, and the increased incomes of the rural residents because it is impossible to solve this pollution only by end-of-pipe treatment, given the experience in treating industrial and municipal pollution.

The second is to turn waste, such as stalk, garbage, and waste water, into resources. Those waste is both pollutants and resources; and the rural residents are both polluters and resource owners. We should work from this angle and allow them to benefit from the resource-using process and at the same time fulfill their responsibilities for environmental protection, instead of replicating the industrial and urban pollution treatment model in rural areas. I believe that we will explore more and better approaches to agricultural nonpoint source pollution by taking the above two pathways.

Another important issue I talked about is the water ecological conservation. We have been way behind the conservation these years and should pay adequate attention to this.

Southern Weekend: My question is about the protection of drinking water sources. We found in previous interviews that several local governments may cancel drinking water source protected areas in order to attract investment projects. What are the regulations about the cancel of drinking water protected areas? Is there any way to prevent this?

Zhang Bo: The circumstances at community levels are complicated. First, laws stipulate that only provincial governments have the authority to approve the plans for delineating and redefining the drinking water source protected areas, which can't be canceled without approval.

Second, some of the drinking water source protected areas need to be redefined, but they have to follow legal procedures. For example, the boundaries for some protected areas were not delineated in a science-based way and cover the places that should not be within the boundary.

Third, there are strict legal provisions on the drinking water source protected areas. It is absolutely prohibited to redefine the drinking water source protected areas with a view to evading legal restraint and giving green light to polluting enterprises. We welcome the public and media to supervise such violations and report to us once they are spotted.

Nanfang Metropolis Daily: Where are we with the mechanism for coordinating water pollution prevention and control between regions, and what measures will be taken in the next step?

Zhang Bo: We have actively promoted this coordination mechanism for water pollution prevention and control and established such mechanism in, for example, the Beijing–Tianjin–Hebei region, the Yangtze River Delta, the Pearl River Delta, and the Yangtze Economic Belt. This is to provide a platform for addressing water pollution across administrative boundaries. It is a challenging task, and the crux is to clearly define the responsibilities, coordinate and supervise efficiently. We are trying in this direction and will keep you informed of the latest developments.

Legal Daily: Some people believe water pollution is as serious as air pollution. What do you think of this? MEP launched a survey on the chemical enterprises along major river banks in 2006. What is the current situation now? Where are we with the amendment to *Water Pollution Prevention and Control Law*?

Zhang Bo: You shared my concern. In spite of all the talks about air pollution, we can't afford to overlook the water pollution prevention and control. The experience in the developed countries suggest that it is more difficult to treat watershed pollution, so we should attach great importance to it. The CPC Central Committee and the State Council have done so and introduced many important measures such as the *Action Plan on Water*.

To properly arrange the distribution of chemical enterprises along major river banks, we should keep in mind the carrying capacity of watersheds and follow the ideas proposed by President Xi that the development of the Yangtze Economic Belt should be facilitated by promoting well-coordinated environmental conservation and avoiding excessive development and that mountains, rivers, forests, farmlands, lakes, and grasslands are a community of life. The fundamental reason for the unreasonable industrial layout and serious ecological damage in some local areas is that they have not well established the awareness of environmental carrying capacity. The environmental carrying capacity varies from watershed to watershed. The environmentally sensitive riverside areas and wetlands that have very high ecological values should be preserved by ecological conservation red lines. However, without such awareness, the wetlands will be undervalued as aquatic plants and reclaimed as industrial zones and built areas to pursue economic benefits. Therefore, it is imperative to keep in mind the environmental carrying capacity and pursue environmental protection of watersheds in line with the important ideas of President Xi and the requirements of the CPC Central Committee and the State Council on ecological civilization development.

In addition, we should delineate and strictly observe the ecological conservation red lines. The local areas should, in accordance with relevant regulations, draw ecological conservation red lines in ecologically sensitive areas with high ecological values and crack down on activities that damage the environment. Lake reclamation is one of the priorities of the ongoing central environmental inspection. Moreover, some areas in the upper reaches of rivers, in order to meet high water demand, build too many reservoirs and hydropower projects, to the extent that the water levels of some major rivers and lakes are not high enough to meet ecological conservation purposes. For example, a survey suggested that 227 out of the 333 rivers captured by the satellite in the Beijing–Tianjin–Hebei region dried up during the flood season last year, not because of lack of rainwater but of water diversion projects. Local areas should ensure that a certain percentage of water resources is allocated for ecological conservation purposes in both water abundance and shortage circumstances, so as to support the aquatic life in rivers, lakes, and wetlands and sustain the self-cleaning capacity of the environment. Otherwise, the ecological damages will undermine the environment carrying capacity and invite the retaliation of nature such as the outbreak of algal blooms. Thus, it is very important to delineate and keep to ecological conservation red lines.

We will take these factors into account while making plans and highlight law enforcement and accountability for water ecological damages.

Water Pollution Prevention and Control Law has gone through the first reading by the National People's Congress and will go through the second reading in June. We will consult all stakeholders between the readings. Thank you.

Ifeng.com: My question is about the waste water treatment in the rural areas. Waste water treatment infrastructure in urban areas is well established, while the rural areas face the common problem of serious pollution caused by discharge of waste water. Are there any measures taken to fix this? How do we prevent and control water pollution for the south-to-north water diversion projects?

Zhang Bo: You asked two questions. We should neither adopt a "one-size-fit-all" solution to the waste water treatment in rural areas nor replicate the practices in treating industrial and urban pollution. We should approach this problem based on local circumstances. For example, for those villages that have grown into the size of a town or become a village-in-city, or have a large population or massive tourists, waste water can be collected and treated like in urban areas. For villages where large numbers of villagers move to cities for a job and the traffic infrastructure is poor, we should consider local conditions in waste water treatment. If human feces and urine as organic matters can be separated from waste water and used for agriculture, the remaining part would hardly develop into runoffs. Even such run-offs appear, they can be treated by constructed wetlands. I recently paid a visit to Liangzi Lake in Hubei province and found some good local experience. A local village has a pond of over 1000 square meters where densely aquatic plants grow. Villagers would do washing besides this pond and discharge waste water directly into the pond. However, the aquatic plants had very strong self-cleaning capability. The water transparency reading is estimated at about 1.5 meters by visual inspection. This shows that the waste water without human feces and urine can be treated in a very simple way. This is why environmental protection in rural areas should be pursued based on local situations.

The water resources are diverted through exclusive channels protected by buffer areas in the south-to-north water diversion project. So the problems you talked about may happen in the headwaters of the project, for example, the nonpoint source pollution in Danjiangkou Reservoir. There is a proven practice called waterfront organic agriculture at community level. We may well learn some useful experience from the community levels. Thank you.

Guangming Daily: The priorities and difficulties in waste water treatment in urban areas differ from those of rural areas. What are they respectively, and how do we balance them?

Zhang Bo: The crux of water pollution prevention and control in both urban and rural areas lies in the recycling of waste water, since the application of only end-of-pipe treatment often gets half the results with double efforts. The older ones with us may recall that the cities used to have cleaners who take feces to the countryside, while now we have waste water pipes that carry waste water to treatment plants. The sludge, in spite of abundant nutrients, may contain heavy metals and thus cannot be applied in farmlands, which cuts off the circulation chain through which human wastes can be applied in agriculture. It is the same case with reclaimed water which has high level of transparency and may well be used for gardening. However, the use of reclaimed water is not so easy. Hundreds of thousands or even millions of tons of waste water are collected for processing daily in waste water treatment plants and it is hard to distribute them again after treatment. Thus, the treatment plants should be more sparsely distributed with smaller treatment capacities in order to recycle the reclaimed water. This will greatly reduce the needs for long-distance water diversion.

To this end, the comprehensive water price should be raised properly. Some local governments may have concerns that the rise in water price may affect the life of lower income earners. This will not happen if we put in place proper measures. The water demands of lower income earners are within a small range, for which the water price can stay unchanged. More important is the water-intensive industries, for which the water price should be raised appropriately to curb their expansion, stimulate the water recycling market and guarantee the amount of ecological flow for ecological purposes.

Water pollution prevention and control in urban areas involves a wide range of aspects. If we head towards the right direction, there will be a bright future.

Liu Youbin: I will call an end to this press conference. Thank you for your participation.

Regular Press Conference (April)

April 21, 2017

On the morning of April 21, MEP held the April press conference, introducing the national environmental supervision and law enforcement situation in 2016 and the 1st quarter in 2017. Tian Weiyong, Director General of the Bureau of Environmental Supervision of MEP, attended the press conference, shared related information, and answered questions from reporters. For the first time, foreign media, including Reuters and CNN, were invited to attend this press conference.

Ministry of Environmental Protection of the People's Republic of China,
2017 Press Conference Records of Ministry of Environmental Protection, the People's Republic of China, https://doi.org/10.1007/978-981-13-7330-5_5

MEP April Press Conference (1)

MEP April Press Conference (2)

Liu Youbin (moderator): Dear friends from the press, good morning! Welcome everyone to the MEP April Press Conference.

In April 2014, the Standing Committee of the National People's Congress examined and adopted the newly revised *Environmental Protection Law*, which is an important part of the rule of law and a major measure to strengthen environmental protection and promote ecological progress. Since the implementation of the *Environmental Protection Law* on January 1, 2015, it has received great attention and expectation from all walks of life. Tian Weiyong, Director General of the

Bureau of Environmental Supervision of MEP, is invited to attend today's press conference to introduce the implementation progress of the *Environmental Protection Law*, especially the national environmental supervision and law enforcement in 2016 and the first quarter of 2017, and answer questions and concerns from you.

Let's welcome Director Tian Weiyong to share with us some information.

Tian Weiyong: Dear friends from the press, good morning! Thank you for your consistent concern and support for the environmental protection progress, especially the environmental law enforcement effort in China. Considering the topic of today's press conference, we will mainly focus on the implementation of the *Environmental Protection Law* in 2016 and the first quarter of 2017.

Before that, I would like to give you a brief introduction of our environmental law supervision and enforcement approaches.

We are the Environmental Supervision Bureau in the MEP. The corresponding departments in local government can be called as Bureaus, Units, Corps, and Battalions. These groups, entrusted by environmental protection departments at various levels, exclusively focus on on-site law enforcement, including on-site enforcement, supervision, inspection, regional dispute mediation, and charge collection. On-site law enforcement is definitely the most essential duty, which includes on-site inspection, administrative compulsion, and administrative penalties.

Tian Weiyong, Director General of the Bureau of Environmental Supervision of MEP

Under the strong leadership of the party group of MEP and the Minister himself, it is fair to say that there has been significant progress in environmental law enforcement and supervision in the recent years. In summary, the achievements we have made in law enforcement can be categorized as the following 4 aspects.

First, constantly strengthened the enforcement.
Second, constantly enriched enforcement means.
Third, constantly created new enforcement measures.
Fourth, constantly improved enforcement capabilities.

Specifically, they include various efforts:

First, constantly strengthened the enforcement. Both the number of cases and the volume of fines set a new record in 2016. In 2016, 124,700 decisions on administrative penalty were issued with the total fine of 6.633 billion RMB. It is a significant leap from 2015: the volume of cases increased by 28%, and the fine by 56%.

From 2015, the newly adopted *Environmental Protection Law* has equipped us with new means, including penalty by days, suspension on production, limit on production, and transfer for detention, which also led to an increase of cases.

Second, continuously enriched enforcement means. In the past, administrative means have been used frequently. Since 2015, the *Environmental Protection Law* approved several new means, which have been constantly utilized, emphasizing on the use of administrative means including admonitory talks, public supervision, and limited approval in regions. These means have all been widely applied in the recent years. This is the updates on administrative means.

In the past years, we have been exploring more possibilities to protect the environment through criminal and civil approaches. The judicial interpretation of the "Two Supremes" (the Supreme People's Court and the Supreme People's Procuratorate) in 2013 and the new version in 2016 were widely adopted as well, which made great achievements in combating various crimes. In addition, we encouraged public participation and public interest litigation for compensation from unscrupulous companies to promote effective law compliance of companies.

Third, continuously created new enforcement measures. In recent years, we used numerous modern law enforcement measures. Take online remote monitoring as an example. We realized 24-h constant monitoring, which means the environmental protection departments can get real-time reports on any issues. Recently, as you may have noticed, we established a program of 14 grids of hotspots. In the past, law enforcement could be a tough job because it was very hard to locate the criminals, or say "enemies", against such a grand area. Through the grid of hotspots, however, we could identify the areas with the biggest emissions, divide the entire Beijing–Tianjin–Hebei area by a grid of 3 km × 3 km through satellite retrieval technology. A total of 36,000 squares has been identified in the grid. Then, we sorted the squares according to their emissions, from the highest to the lowest. 2,970 squares were selected, which accounted for about 80% of the total emission. These are our main focus and we could locate key objects subject to supervision within these areas. We have also adopted IT-based measures to keep law enforcement records and tracks.

Fourth, constantly improved enforcement capabilities. Through a rotational training for 3 years, all law enforcement officers were "fully covered". Training constantly strengthened law enforcement standard.

In addition, "Two Offices" (the General Office of the CPC Central Committee and the General Office of State Council) issued notices on the direct oversight of provincial-level environmental protection agencies to their respective local departments, which set solid foundations for law enforcement positioning, vehicle supplies, and attire requirements, etc. All these factors are significantly beneficial for the law enforcement.

Now, I'm ready to answer your questions if there is any.

Liu Youbin (moderator): We'll open the floor to questions.

China National Radio: the undergoing inspection on "2+26" cities in the Beijing–Tianjin–Hebei Region is the biggest one in the history. Could you please give us some updates? What are the major issues identified during the inspection? How to effectively solve these problems?

Tian Weiyong: The year-long inspection of "2+26" cities in the Beijing–Tianjin–Hebei Region you just mentioned is indeed a major action we are currently taking. We ran this inspection once in the entire month of February, with proved result in better air quality. It is expected that we can realize some existing goals by constantly strengthening the inspection. In this year's Report on the Work of the Government, Premier Li Keqiang proposed to defend our blue sky and to conduct special campaigns to address pollutions in key industries. It is required to implement 24-h online high-quality monitoring on all pollution sources in key industries. For the companies that cannot meet the standards within the required time frame, it is necessary to firmly shut them down according to the law. This is a solemn commitment made by Premier Li to all people nation-wide during the two sessions.

In order to meet the commitment, the leading party group of the MEP made a major decision on carrying out a one-year project focusing on defending the blue sky, which is also the core task of this grand inspection.

The inspection began on April 5. Entrusted by Minister Chen Jining, Zhai Qing, Vice Minister of the MEP, made a national mobilization. We selected 5,600 people country-wide and implemented 25 rounds of inspection. For each round, 28 cities were inspected with 8 inspectors for each city. The transitions between every two rounds are very important. The first round took 2 weeks, so did the second one. If there were no transition, then the second group could not know what the first group did. Therefore, we decided to have people from 2 groups to work simultaneously during the handover week to ensure the effective cohesion between groups. From the first group to the second, and from the second to the third.

Today is the 21st, which coincides with the official launch of the inspection for 2 weeks starting from the 7th. The week between the 21st and the 28th is for the handover between the first and the second group. It is important to clarify a lot of things: what was inspected? What issues were found? What was handed over to local governments? How did they implement? The second group needs to have all the related information. The second group then have a foundation set by the first group, instead of starting from scratch.

The handover between the second and the third group will just be the same. Half of the first group will stay until the 28th. Afterwards, the second group will fully take over. A total of 25 rounds of inspections will be conducted with this approach for one year.

By 24:00 yesterday, 28 groups inspected 4,077 companies from various industries, and found 2,808 companies, 69% of the total, with different environmental problems. According to the original plan, the focus in April is to inspect the companies that are "in unauthorized locations, lacked the proper certificates and failed to meet emissions standards". During this inspection, 763 companies were found problematic, accounted for 27%. In addition, 512 companies had no equipment for pollution control or had equipment with malfunctions, accounted for 18.2%. Another 15 companies exceeded the emission standard, 8 companies practiced fraud on their automatic monitoring equipment, 687 companies did not adopt dust prevention measures or had faultiness, etc. These are the major issues we discovered during the first round of inspection for 2 weeks, which have been handed over to local governments. Cities have their own lists, or say account books, in order to hold them accountable for each issue. To firmly conduct this task, local authorities need to ensure all issues are solved in time to write them off from the book.

CCTV: We learned that several seepage pits for sewage water were discovered around the North China area. Some have been treated from years ago, but the effect was not obvious. What on earth led to these seepage pits? How hazardous could they be? Has the MEP conducted any related investigation? How many pits are there in total? Why were the treatments not so effective? What specific measures will be adopted to solve these issues in the future? Thank you.

Tian Weiyong: This is a question that received much more attention lately. The MEP got the information on the morning of the 19th and immediately established a working group for on-site inspection. It is true that the facts media brought up do exist, and we made it public to the media instantly. I would like to thank the media for the supervision first. With your effort, we can discover and solve many problems in a timely manner.

The 2 seepage pit incidents in Tianjin and Hebei are currently still under full investigation. However, there are at least two kinds of illegal activities are involved. Firstly, discharging pollutants in seepage pits and wells to escape supervision. No matter how the pits were formed, no matter how long they have existed, they are used to illegally discharge pollutants. Secondly, illegally dumping hazardous wastes. As early as in 2008, when the *Water Pollution Prevention and Control Law* was revised, seepage pits and wells were listed as illegal acts from the legislation perspective, which means there are, indeed, very clear regulations on such acts in the law. In 2013, the judicial interpretation presented by the "Two Supremes" (the Supreme People's Court and the Supreme People's Procuratorate) considered these acts as criminal behaviors that can result in sentences. Since the implementation of the *Environmental Protection Law* in 2015, more means became available. All of the five new measures of the *Environmental Protection Law* can be applied to solve these issues, which shows the severity of these issues requires even more serious measures. The state needs to severely crack down this kind of acts.

The MEP has attached great importance to this incident. First of all, the MEP is organizing a detailed inspection nation-wide and the result will be disclosed to the public. Second, on behalf of the MEP, I would like to say that the MEP will severely handle each of the cases we found out and never tolerate such acts. Third, we welcome the media and public to supervise and report on such issues. Some of such issues are located in very remote places. The "12369" WeChat account is available for receiving reports. It has 240,000 followers and I hope all of you can call for more attention to it. People can report, pay attention to, and support the environmental protection effort in China through the WeChat account, to bring all illegal and criminal acts to light. Thank you very much!

Reuters: Good morning. The seepage pit issues were initially brought up by an NGO based in Chongqing. We noticed that after the media coverage, local environmental bureaus executed fast response, which was very impressive. Based on the experiences in the past, the local government may encounter some difficulties during environmental protection supervision and law enforcement. I would like to know what is the government's perspective on the role of NGOs? What kind of influence should they have?

Tian Weiyong: We absolutely welcome the inspection on environmental-related issues from all kinds of NGOs, the public, and media. We encourage you to disclose such issues to the society or report them to the environmental protection departments on various levels. This is, firstly, the attitude of the MEP.

Secondly, we hope that more and more NGOs can support and pay more attention to the solution of environmental issues. As I mentioned earlier, we hope that we can better leverage the supervision of NGOs in promoting company's compliance. In addition, public interest litigation is another kind of supervision we should utilize to promote compliance. Through the joint effort of the government, NGOs, and media, the environmental protection issues will be solved in short.

Liu Youbin: I would like to add something. NGO-related issues are the responsibility of the Department of Education and Communications. The leading party group of the MEP has attached great importance to this issue.

First, together with the Ministry of Civil Affairs, we printed and distributed the *Guiding Opinions on Strengthening the Development Guidance and the Standardized Management of Environmental Protection Civil Societies* to the national environmental protection and civil affairs systems. That was the first time for us to issue a document about further promoting the effort on environmental protection civil societies and fully leverage their essential position and influence together with the department of civil affairs. The core intention of this document is to remind local environmental protection and civil affairs departments of the importance of environmental protection NGOs.

Second, we organize trainings for environmental protection NGOs every year. For example, we just organized one for NGOs nation-wide in Guangzhou and some environmental NGOs participated in that training session.

Third, we provide some small financial aid every year. We provided funds for several NGOs last year and we will continue to do so this year.

Fourth, we will try to build up a communication platform between the government and the NGOs and hold seminars and meetings regularly. Earlier this year, we invited a dozen of representative NGOs in Beijing for a seminar. The seepage pit issue was initially discovered by environmental protection civil societies and the related governmental departments took necessary measures and responded quickly.

Caixin Media: On April 4th, Mr. Chen Jining, Minister of the MEP, led a team and investigated Yanshan Petrochemical, found out that their VOCs were not effectively managed. What are the specific issues of their VOC management equipment? Are they common among petrochemical companies? What kind of obstacles have you encountered during the inspections on VOCs? How to solve them?

Tian Weiyong: Minister Chen did find out some issues on the VOC management during his on-site investigation of Yangshan Petrochemical. The 3 sets of auxiliary equipment for watering and desizing in the rubber workshop had some design defects in its process. VOCs were not efficiently treated after getting recycled into the system. It caused secondary pollution due to volatilization. Currently, this set of system has been modified with the intervention and support of Sinopec, and the problem has been solved rather well.

It is fair to say that VOC management issue has always been a prominent one in the petrochemical industry. The Yanshan Petrochemical incident disclosed an aspect of this problem. There are several obstacles. For one thing, the workload for VOCs supervision is quite heavy. Take Yanshan Petrochemical as an example: during my last investigation, there are more than 800,000 possible leaks, which need to be examined and checked one by one.

VOC is also an issue related with other industries. Painting, package printing, and other processes can all generate VOCs, even the agricultural industry can generate VOC issues. To solve these issues, the MEP will take the following measures.

First, strengthen the control from the source, restrict the emission of VOCs of companies from different industries. To reach this goal, we need to focus on the integral control.

Second, from several points to the whole picture. It means that starting from the industries with high VOC emissions, including petrochemical, organic chemical, package printing, we will strengthen the control and management first.

Third, prioritize the effort in controlling the key points, including the sealing points of equipment, storage, and loading and unloading.

Forth, improve the competence and efficiency of law enforcement. We will use some portable devices for law enforcement to strengthen the management and control. In the past, we identify issues through its smell. However, the accuracy cannot always be guaranteed, which means we need to rely on devices and equipment for a better control and management.

Fifth, promote the effort in disclosing information to the public and encourage public participation, to form a virtuous cycle between supervisors and supervisees.

Southern Weekly: My question is about the standard for selecting the personnel for this inspection on air pollution prevention and control. By transferring inspectors from other places to the Beijing–Tianjin–Hebei area, will it cause problems for other regions due to the short of staff? You mentioned that the MEP has investigated and panelized some companies during the past 2 weeks. Will this inspection focus on inspecting individual companies? Or leveraging the inspection on companies' illegal acts of environmental pollution as an instrument to drive the government to shoulder the overall responsibility of controlling environmental pollution?

Tian Weiyong: This time, we mainly selected inspectors from the Beijing–Tianjin–Hebei area internally, instead of transferring a lot staff from other regions. Among the 5000 law enforcers, 3000 were from local Beijing-Tianjin-Hebei area. The rest of 2000 are from other provinces. Among the staff, some were recommended by the provincial government, who are very experienced in law enforcement as leaders or deputy leaders of corps. Leaders of corps are the city-level inspector leaders. They lead a group for the inspection. Second, we required that each group should have a deep understanding and related background matching the industrial features and characteristics of their target cities. For example, if the person dispatched is very experienced in inspecting paper production industry, it might not be a good idea to send him/her to a city that heavily relies on the steel and iron industry. That means we need to assign the people with matching professional background to specific places so that they can start their work fairly quickly.

In addition, it may have a rather big driving force to the area. The groups sent to Zhejiang and Hangzhou did a great amount of work in anti-counterfeiting. They were the major force in that effort. This is how the members were carefully selected and assigned according to the law enforcement requirements in each specific area, targeting its unique issues.

To answer your second question, this inspection is a combined effort focusing on both the government and company capabilities. There are mainly 7 tasks in this inspection. First, urge governments at various levels, especially the ones at city and county level, to fulfill the requirement raised by the Beijing–Tianjin–Hebei pollution prevention and control program while keep tracking their implementation progress. This is the focus of this inspection.

If the governments of various levels cannot complete all tasks, they will need to take measures and shoulder the responsibility. There is a list of tasks for the governments that clearly defines the responsible persons for each task. They are all government staff or from government-related departments.

Besides, you have noticed our aggressive inspection on companies. The focus of this inspection is the monitoring feedback of pollution sources high above the ground. In addition, pollution treatment facilities and the emission status of major pollutants of stationary pollution sources are our focus as well. One of the key goals we are striving for this year is to ensure that big enterprises can meet the emission standards steadily.

As for the many companies in the Beijing–Tianjin–Hebei area that are "in unauthorized locations, lacked the proper certificates and failed to meet emissions standards", this inspection required the "2+26" cities to run self-investigation before the end of March. They carried out a list of 56,000 small companies with "location, certificate, and emission" issues.

These companies need to be culled out one-by-one by this October, no one shall be left behind.

For the companies that cannot conduct production during the heavily polluted winter times, they can produce more during the days with better air quality, and stop producing when it is bad. It is also very important to execute this requirement of off-peak production, but we need the companies to carry out plans and report them to the environmental protection authorities. Then there is governance of VOCs as well.

This inspection is focusing on both the government and companies. It is a combination of both aspects.

The Beijing News: You mentioned that there are quite a few companies that are "in unauthorized locations, lacked the proper certificates and failed to meet emissions standards" in the Beijing–Tianjin–Hebei area. However, after all, inspection personnel are limited. How can we make sure that this kind of issues will not happen again after this one-year inspection?

Tian Weiyong: How to consolidate the inspection result is definitely a key question. When we were planning the inspection, it was decided that the inspection should not replace local supervisions. Issues found out during the inspection should be handed over to local governments, and kept track of the progress. How to assess if the issues are solved? There is going to be follow-up groups to take care of the transition work and review if the former group has completed its tasks.

In addition, we will dispatch follow-up inspection teams to review the discovered problems and track the progress from time to time. We have 5,600 people as inspectors and additional follow-up inspection teams. Let me tell you something, we have also selected a dozen of invincible experts who are experienced in handling cases. They will occasionally visit different sites for supervision. It is important to monitor the implementation of solving discovered issues, as well as to identify new ones. Both aspects will define the performance of our inspection groups. Issues can be effectively solved through the effort from multiple aspects and approaches.

The Minister of the MEP once said that he will lead a team and go for an inspection himself. If the issues cannot be solved properly or rebounded, then the local government should be held accountable. We can take measures of having admonitory talks, or restricting their project approvals, including calling related persons to shoulder the responsibility according to the ecological damage. Truly implementing these measures can ensure that all issues will be solves.

China Daily: It is reported that during the inspection, a lot of illegal acts have been discovered, while the companies refused to cooperate. How do you solve this problem? How to guarantee the effectiveness of law enforcement? Or the safety of law enforcers? It is said that some cities now have environmental protection police officers, what are their responsibilities? How can they cooperate with the environmental protection department? Thank you.

Tian Weiyong: We did encounter some obstacles of law enforcement during the inspection. We had the following difficulties. One is obstructing law enforcement. You may have heard that companies in Jinan and Xingtai kept our inspectors for more than 1 h. Others may be refusing to be inspected. Companies might shut their doors once they saw inspection groups and refused to open it no matter how hard the inspector tried. They stopped production as well. It is quite common.

Another severe reaction is violent defiance against law enforcement. We adopted different measures to handle the aforementioned issues. For refusing inspections, we can fine the company according to related regulations. In terms of obstructing law enforcement, we can cooperate with the public security department for administrative detention for 5–10 days. There are clear terms stated in the *Public Security Administration Punishment Law*. Violent defiance against law enforcement is a serious crime which is subject to sentences of imprisonment for up to three years.

These standards are already established. They need to be executed accordingly against different types of violations. There are many police officers in the public security department. With their support, our environmental protection law enforcement is more effective.

The measures I mentioned earlier, including the ones covered in criminal law, cannot be implemented without the support from the public security department. The environmental protection department will transfer the issues discovered to the public security department to file the case and conduct further inspection and law enforcement, including taking the measures like administrative detention and investigating for criminal responsibilities. With the support of the public security department, our environmental protection law enforcement is now much more effective.

Especially in the last year, together with the public security department and the Supreme People's Procuratorate, we issued the measures connecting the "Two Laws", which efficiently connect administrative law enforcement with criminal justice. Transfer and intake the cases needed to ensure an effective and promptly handling of illegal acts and crimes.

Beijing Youth Daily: Last September, the General Office of the CPC Central Committee and the General Office of the State Council printed and distributed the *Guiding Opinions on the Pilot Project of Reforming the Vertical Management System of Monitoring, Supervision and Law Enforcement of the Environmental Protection Institutions Below Provincial Level* (hereinafter referred to as the "*Opinions*"), is there any updates about this reform?

Tian Weiyong: This is truly a big reform plan that covers several sectors including the monitoring, supervision, and law enforcement of environmental protection institutions. The "Opinions" is a strong support to our law enforcement. I would like to brief you from the following aspects.

First, it was clarified that environmental protection law enforcement should be included in governmental administrative law enforcement. This had been a goal our environmental protection department striven for but could not achieve. Documents issued by the central government specified that environmental protection law enforcement should be included in administrative enforcement.

Second, it empowered environmental protection law enforcement with the right to conduct on-site inspection, administrative compulsion, administrative penalties, etc. The "Opinions" clearly stated that it is important to empower related departments to better facilitate the law enforcement effort.

Third, it specified our attire. It is quite exciting for our environmental protection department. The uniform has been put on and taken off several times. You mentioned issues of obstructing law enforcement, which was partly attributed to the lack of uniform. People recognize uniforms as representing official troops. When we were inspecting companies before, everyone just put on their own clothes without a unified attire. No one took us seriously.

The "Opinions" also pointed out to guarantee the supply of vehicles for law enforcement. The enforcement needs to cover a wide range and scope of objects. Especially in the mid-western area, it is extremely difficult to get to the site on time without vehicles. So it is very important to guarantee vehicle supply. The "Opinions" also clarified the management of personnel quality and the standardization of inspection groups.

According to the guiding opinions from the CPC Central Committee, step by step, we are implementing every item in detail with specified plans. By implementing the "Opinions", the capability of the law enforcement group will develop to a higher level.

Jiemian.com: We noticed that some companies just turn their environmental protection equipment on during the inspection, turn them off and directly emit wastes again when inspectors are gone. Earlier when the Minister led on-site inspection, there was no prior reminders sent. Will there be any reminders this time? How can we radically change companies' mindset of being forced to meet the standard instead of proactively to do so? Thank you.

Tian Weiyong: The normality of company compliance has always been a priority of environmental protection law enforcement. Illegal acts of companies and substandard emission are indeed quite common. Statistics show that during the 2 years of implementing the *Environmental Protection Law*, online monitoring devices are installed nation-wide in the key companies requiring emission supervision, to replace the original inspection approach that sampling emission wastes for comparison and analysis. Here is a graph I would like to share with you. In the 1st quarter of 2015, only 53.1% of companies met the standard. Since then, the rate gradually increased to 71% by the 1st quarter of this year. The standard fulfillment rate of the key pollution sources under national control gradually improved.

If we count in the companies that exceeded the standard by 10%, the rate is even higher. It increased from 78% in the 1st quarter of 2015 to 90% in the 1st quarter this year. Through thoroughly implementing the *Environmental Protection Law* in the recent years, while leveraging the aforementioned judicial and administrative measures to enable the normality of company compliance, we have witnessed a gradual improvement in the standard fulfillment rate.

Especially in the Beijing–Tianjin–Hebei area, where, as you just mentioned, is under the spotlight of public attention, the statistics from our survey show that between January and the end of December last year, the rate of exceeding standards decreased continuously, from 31% at the beginning of last year, to 3.79% at the end of December, which means that the standard fulfillment rate increased from 69% to 96%. It is shown that the standard fulfillment rate in the Beijing–Tianjin–Hebei area is higher than the average national rate. Overall, the standard fulfillment rate has been improving gradually.

In order to minimize the operational cost, some companies chose to illegally discharge wastes while evading regulations. We adopted many countermeasures as well. Firstly, we adopted online remote monitoring targeting big enterprises. In addition, we need to inspect the smaller ones more frequently. We do not need to conduct on-site inspection on big companies all the time as remote monitoring is available. If it is had not been adopted, we would need to inspect them much more often constantly. The reason for us to conduct this inspection is to continuously check on these companies, sparing no time for them to catch a breath. By continuously doing so, issues with illegal acts will definitely get mitigated.

People's Daily: During the intensified inspection recently and the prior law enforcement, some companies were found out practicing fraud on the monitoring system of pollutant discharging. What measures can be adopted to solve this problem?

Tian Weiyong: The issue of practicing fraud on the pollutant source has always been our focus, especially manipulating the online monitoring devices. We need to crack down on these practices. You may think the rate of companies practicing fraud is high. Let me show you some numbers. There are 2,808 companies found problematic in the inspection on the Beijing–Tianjin–Hebei area up till yesterday, in which 8 companies practiced fraud, taking 0.3% of all the companies conducted illegal acts. It is fair to say that they only took a small percentage. Even though it only counts as a very small part, it is very difficult to solve the problem since it requires professional knowledge and technology.

There may be 3 loopholes where the company can practice fraud on. Firstly, sampling. It is easy to practice fraud on this step, like substandard sampling or missing some samples. Second, analyzing. There may be some problems with calibrating. Third, uploading. As this step is conducted remotely, companies may set some limit on the data by themselves and simply stop uploading the data that exceeds the limit. The practices are changing all the time as well.

The next step is to encourage all companies to install online monitoring systems. If there is one edgy advantage of environmental supervision in China comparing with the world, it is our online technology. We have installed online monitoring system in more than 10,000 SOEs, which serves as sentries who can identify problems promptly. Anti-fraud is another step. We are going to launch dedicated anti-fraud campaign, establish related measures, publish a booklet to list out all of the possible scenarios and practices. Through statistical analysis, we can tell if there are any fraud practices tentatively. There are also many controlling measures we can take to stop those practices. But more importantly, we need to rely on public supervision. Statistics should all be publicly available. We have initiated some pilot projects in several areas. For example, we are sampling water regularly and keeping them stored. There is a sample at 3:00, and we can see the discharging data through the monitoring system. Afterward, we will take the sample for an experiment to see if the lab result matches the original one. All of these measures are adopted to ensure the stability of online data.

China News: The quarterly growth rate of illegal cases was quite high. For example, the year-on-year growth rate of the number of growth rate reached 93% in 2016. What is the reason? Was it because that companies decided to take a chance and refused to install environmental protection equipment? Or was it because that the law enforcement effort got stricter?

Tian Weiyong: You probably noticed that China University of Politics and Law conducted an assessment on the implementation of the *Environmental Protection Law*, which was quite an objective reflection. In addition, Renmin University also assessed 4 auxiliary measures of the Law, which are detailed approaches related to the implementation. Last year, Morgan Stanley from the US also evaluated the implementation of the *Environmental Protection Law* and published the result.

This issue has drawn tremendous public attention. I deliberately downloaded several sentences from the assessment done by China University of Politics and Law yesterday. The first sentence, local party and government leaders paid unprecedented attention to the environmental protection work. The second one, the law enforcement effort on environmental protection supervision improved significantly. Third, people from all walks of life have paid more attention to and been more actively involved in the environmental protection effort. These are the 3 main conclusions. In addition, it also emphasized that the implementation, compliance, and influences of the stated systems and measures of the *Environmental Protection Law* have been unprecedented in the history of environmental protection.

I would like to talk about the reasons for the *Environmental Protection Law* getting such a comment. From my personal perspective and work experience, I have the following points.

First, the *Environmental Protection Law* has been more deeply implemented. From January 1st when the new *Environmental Protection Law* was officially implemented, we have organized 3 implementation-year campaigns successfully. This year, the implementation campaign specifically required all counties, cities, and districts to comprehensively adopt all detailed measures and means stated in the *Environmental Protection Law*. Last year, it just required all local cities to adopt regulations related to the 5 types of cases (penalty by days, sealing up and distraining, cutting or restraining production, transferring for detention, and transferring as suspected crimes). This year, the requirement was to all counties and districts, which guarantees the implementation of the *Environmental Protection Law*.

Second, cases of various types increased dramatically every year. There were 22,730 cases of the 5 types last year, increased by 93%. In the 1st quarter this year, it increased even faster, reaching 195% comparing with the same period of last year. Last year, there were 9,976 cases adopted sealing up and distraining measures, increased by 138%. 5,673 cases about cutting or restraining production, which increased by 83%. 4,041 cases of transfer for detention, grew by 94%. 2,023 cases were transferred as suspected crimes, increased by 20%. It is shown that various types of cases have increased significantly.

Third, same as last year and the year before, the bounce rate of illegal acts maintained at a low level. The bounce means when an illegal act was caught of happening again. The bounce rate was 2.86% in 2016, and 2.57% in 2015.

Fourth, the waste discharge standard fulfillment rate has been gradually improving. Just as the 2 graphs I just showed, the rate increased from 78 to 90%, and from 69 to 96% in the Beijing–Tianjin–Hebei area, both can prove the improvement.

Environmental quality and law enforcement measures are closely coordinated. Last year, we pointed out 5 cities that either did not meet the environmental quality standard, or had deteriorated index, or adopted very few law enforcement measures. We are going to promote supervision effort on these issues to help solve different types of illegal acts in this area. Comprehensively implement the *Environmental Protection Law* through the aforementioned measures.

Legal Daily: The assessment conducted by the Law School of Renmin University mentioned that the implementation rate of penalty by days was only slightly over 40%, is that true? How much of an influence can administrative detention have? I noticed that comparing with 2015 and 2016, the number of administrative detention cases doubled last year. Is this measure truly effective? Besides, has it ever been the case that the responsible person of a company asked one of his/her employees to replace him for the detention?

Tian Weiyong: Penalty by days has the lowest applicable rate among the 5 types. As I just mentioned, in the 22,730 cases of the 5 types, 1,017 cases were about penalty by days. During the implementation last year, local governments provided input to this issue and touched upon some defects of our system. This regulation was borrowed from laws in the US. According to the regulations in the US, if there are some illegal acts found in a company or its discharge cannot meet the standard, usually, it is needed to trace back to the previous inspection and treat the period in between as breaching the law, unless the company can self-prove its compliance. There were some revisions in the *Environmental Protection Law* and clarified that when there are illegal acts, an order should be sent first pressing for rectification. Afterwards, inspection will be adopted to check if any rectification has been implemented. If not, penalty by days will be followed. Minister Chen mentioned this issue during a Q&A session with the media. It can be said that there is an essential defect in this regulation, which led to low applicability.

Among the 5 types of cases I covered earlier, cases of penalty by days had the highest bounce rate, which reached 9.81%. After the first round of penalty against companies, we can still find issues during inspections and penalty by days will be adopted again. It shows that this measure has weak deterrent effect. Other measures, such as administrative detention, the bounce rate was only 1.4% something, lower than 1.5%, which means less than 1.5% people who got detained twice.

But the bounce rate for penalty by days reached 10%. It reflects that there are some flaws in the mechanism of penalty by days that need to be improved. We are continuously promoting the mechanism design.

In terms of administrative detention, direct responsible persons and executives can be detained according to regulations. I cannot think of a specific case right now, but I am sure the issues of employees replacing the responsible person for the detention you mentioned had happened, and the core solution is to identify the right person. I understand your point, the big boss may have instigated someone to conduct illegal acts and that person got detained or sentenced. The boss could just provide some compensation and shoulder no responsibility. There have been such cases but we have not found any. The problem is how to find the mastermind behind. It is hard to trace back to the mastermind without any evidence. In fact, we need to solve this issue from the legal perspective and clearly define if the boss of a company need to shoulder joint responsibility or relevant liability.

The issue you mentioned is mainly a task for the public security department. Any cases related to administrative detention and criminal offense need to be jointly solved by the environmental protection and public security departments. In the future, environmental protection departments will mainly focus on providing evidences for the public security department for investigation and judgment.

China Youth Daily: There are about 56,000 companies that are "in unauthorized locations, lacked the proper certificates and failed to meet emissions standards". Does it worth it to invest such a huge amount of resources into inspection just to solve the related issues? I remember some information about including some clothes stores as this type of problematic ones in Beijing. Could you please let us know how much did these companies contribute to the pollution? What is the plan for the next step? All of the companies that cannot meet the standard will need to be closed in October as required, why are you still investing so much into this issue? Thank you.

Tian Weiyong: It was a surprise for us when we got the statistics of the companies with "location, certificate, and standard" issues in the Beijing–Tianjin–Hebei area. The statistics pointed out 56,000 companies, and another 700 were discovered in this inspection. What does it mean? We have different interpretation of the standard of this type of companies. After getting the report, in some places, the number was not accurate. We could not understand their purpose, but when we got to the site, there was no company or it had been closed years ago. Another problem is that some may count small clothes stores and mini-marts as companies with "location, certificate, and standard" issues. The interpretation of this regulation is indeed extremely different.

The core task is to accurately pin down the companies with such issues and make a list. Local governments need to identify the ones that need to be shut down or can never meet the standard. Then in October, if problems still exist, that is your responsibility. If there are some new ones discovered that cannot fulfill the requirement and added to the list, that is your responsibility as well. Just give us a list as short as possible is definitely not the end. Some may practice fraud and include some companies that do not even exist to the list in order to brag about their achievement, this is also not acceptable. We are going to strictly investigate on both these two aspects and ensure a significant improvement on the issue of companies with problematic "location, certificate, and standard". It is one of the major measures to solve the air pollution issue of the Beijing–Tianjin–Hebei area, and an essential step to defend the blue sky, which will set a solid foundation for rectifying and prohibiting the production of these companies.

Liu Youbin: That is the end of today's press conference. Thank you for your participation!

Regular Press Conference (May)

May 23, 2017

On the morning of May 23, the MEP held the regular press conference to introduce the *13th Five-Year Plan for National Environmental Protection Standards*. Zou Shoumin, Director General of the Department of Science, Technology, and Standards of MEP, and Wu Xuefang, Director of the Institute of Environmental Standards, Chinese Research Academy of Environmental Sciences, attended the press conference, released relevant information and answered the questions posed by journalists.

Ministry of Environmental Protection of the People's Republic of China,
2017 Press Conference Records of Ministry of Environmental Protection, the People's Republic of China, https://doi.org/10.1007/978-981-13-7330-5_6

MEP May Press Conference (1)

MEP May Press Conference (2)

Liu Youbin (Moderator): Friends from the media, good morning! Welcome to the regular press conference of May.

Environmental standards, as an integral part of national environmental law-making, are the basis for developing environmental protection programs and plans and for the competent environmental protection departments to administer by law. They are of great significance to improve environmental quality, cut pollution emission, facilitate environmental technological and scientific advances, and protect people's health.

At today's press conference, we have with us by Zou Shoumin, Director General of the Department of Science, Technology, and Standards of MEP, and Wu Xuefang, Director of the Institute of Environmental Standards, Chinese Research Academy of Environmental Sciences. They will brief you about China's environmental standards, particularly the 13th Five-Year Plan in this regard, and answer your questions.

Zou Shoumin: Friends from the media, good morning! Environmental standards are important for improving environmental quality, mitigating pollution and preventing risks. They are an important basis for environmental management, supervision and inspection and a major safeguard of people's health. So far, China has promulgated a total of 2,038 national environmental protection standards, 1,753 of which are valid, as well as 167 mandatory local standards on record. A two-tiered framework of environmental protection standards which fall into five categories has been established. The two tiers are national and local, and the categories are environmental quality, pollutant emission (control), environmental monitoring, environmental management specifications and the technical principles and methods on the standard formulation.

The public has long been closely following the leniency and severity of environmental standards and the number of pollutants targeted. There are 16 valid national standards on environmental quality, which cover the main environmental elements such as air, water, soil, noise and vibration, and nuclear and radiation; and 163 valid national standards on pollution emission (control), including 75 air pollutant emission standards that regulate 120 pollutants, and 64 water pollutant discharge standards that regulate 158 pollutants. Overall, in terms of the number of pollutants targeted and the severity of air and water pollutant emission standards, China is on par with major developed countries and regions.

I would love to take your questions. Thank you.

Liu Youbin: The floor is open now.

Southern Metropolis Daily: According to the 13th Five-Year Plan for national environmental protection standards, the soil environment quality standards for agricultural land and construction land will be released by the end of 2017. How will they be different from the current standards? And how will the "category-specific" management requirement be reflected?

Zou Shoumin, Director General of the Department of Science, Technology, and Standards of MEP

Zou Shoumin: The current *Environment Quality Standards for Soils* was issued in 1995 and mainly aimed at preventing soil pollution, protecting the ecological environment, safeguarding agricultural and forest output, and protecting people's health. As it was issued early, the standard does not cover construction land. In recent years, we have issued a series of technical specifications for the surveys, monitoring, evaluation, and restoration of some sites, which act as technical support for treating and restoring contaminated sites.

However, being severely outdated, the standard applies to a small scope and a small number of pollutants, and sets unreasonable limits for certain indicators. Therefore, 3 years ago, we started to accelerate the revision of the standard. A significant change is a structural adjustment, which divides the standard into two parts to regulate the soil environment quality of agricultural lands and construction lands separately. We have canvassed public opinion for three times since 2015 and are still working on the revision.

The new standards are different from the original ones primarily in that they target the characteristics of soil pollution, and are based on the prevailing soil environment risk management and thus reflect the changes of soil environment management. The category-specific management is reflected in several ways. First, the new standard on agricultural lands and that on construction lands are formulated separately; second, the new standards exercise category-specific management based on the severity of contamination and types of land use, and take such measures as restoration, change of land uses, and plantation of alternative crops. Specifically speaking, if a lot is not suitable for growing vegetables, then we can switch to grain crops. If that does not work either, we can try fruit trees. If construction land is not habitable, it can be used for industrial development. If that does not work either, a park can be built on it. Exposure to various pollutants can be controlled in different ways depending on their source to reduce the harm to health. The soil quality standard is different from water and air standards in that it is not all about restoration; it is also about the regulation of land use, which can play a preventive role. We hope to release the two standards by the end of this year.

Financial Times: Will the U.S. withdrawal from the Paris Agreement affect China's future environmental agenda? Will China take a different approach? A lot of coal industries like coal chemical industry and coal power industry can be seen in the northwestern region of the country. How does MEP view this phenomenon?

Zou Shoumin: Your first question is beyond my expertise, but I can try and give a brief answer. China will unswervingly stick to its own development path and implement the *Paris Agreement* as it has pledged. China will fulfil the obligations and requirements of the agreement according to the common but differentiated principle and continue to honor its commitments. The National Development and Reform Commission is in charge of climate change issues. You may ask it for more details on China's involvement in the *Paris Agreement*.

Regarding your second question, MEP is drafting the *Emission Standard for Coal Chemical Industry*. The already published *Emission Standard for Air Pollutants from Thermal Power Plants* is the most rigorous of its kind in the world, and some power companies are still upgrading their facilities to achieve ultra-low emission. We will help optimize industrial layout through environmental impact assessment that is based on social development and local environment. A company decides where to seat its plant as long as it observes the national industrial policies and layout requirements. Environmental protection authorities need to make use of the standards, strictly enforce the law and supervise enterprises' emission to ensure that they meet national environmental requirements.

Jiemian News: It is reported that MEP is carrying out a major project on the cause and control of haze and smog. Are you at liberty to share some outputs? Some industrial insiders argued that the wet-process desulfurization and denitrification contributed to smog to a certain extent. Does MEP agree? What is the share of the dry-process and wet-process desulfurization and denitrification, respectively? How much do the wet-process technologies contribute to smog?

Zou Shoumin: Smog is a common concern. During the two sessions of this year, Premier Li Keqiang expressed his hope of stepping up scientific research to overcome this difficulty. Overall, since the start of the 12th Five-Year Plan period, plenty of national scientific and technological programs, some of which are supported by MEP, have conducted early-stage research on the causes of smog and reached a preliminary consensus on the causes of heavy air pollution, especially in winter: high pollutant emission is the internal cause, adverse meteorological condition the external cause, and more intensive chemical conversion the driving force.

But why are the media and general public not clear about the causes? The reason is the absence of a universally recognized scientific explanation. The study on the smog formation mechanism in detail is not thorough. That is why Premier Li

Keqiang demanded that MEP and competent authorities assemble an outstanding scientific research team to address mainly three issues: first, the causes and formation mechanisms of heavy air pollution, which should be basically unanimously accepted by people from all walks of life; second, a list of pollutants that lead to heavy air pollution and the corresponding management and control technologies; third, protection of people's health during days of heavy air pollution and scientific explanation to the public.

MEP is working with the Ministry of Science and Technology, the Chinese Academy of Sciences, Ministry of Agriculture, China Meteorological Administration, National Health and Family Planning Commission and universities to draft an implementation plan and conduct a preliminary study. I estimate that they will reach initial conclusions during next year's two sessions. But I am sure you all understand that haze and heavy pollution weather do not come into being overnight. To get to know their causes and solutions, we still have a long way to go.

Regarding your second question of whether the wet-process desulfurization aggravates smog, not much research has been done and the scientific community cannot yet reach a consensus, let alone a final conclusion. Thank you.

The Paper: Environmental standards are an integral part of national legal standards. During MEP's law enforcement, it is common that a company reported having polluted the environment is found to meet pollution standards at the time of inspection. Earlier a central environmental protection inspection turned out to be just like this. This is the same case with water pollution. Last year, a water pollution incident was reported in Jiangxi province, however, all water quality indicators were within limits during routine inspections and none of the pollutants causing the incident was spotted. What do you think of this? Second, earlier documents mentioned that during the 13th Five-Year Plan period our national environmental standards are going to be adjusted to align with pollution permit and other new systems. Could you please elaborate on the adjustment and alignment?

Zou Shoumin: Emission and discharge standards are a yardstick against which we can measure a company's compliance with standards. As we often say, exceeding the standards means breaking the law. The existing environmental regulations explicitly stipulate that all polluters shall comply with relevant laws, regulations, requirements, and standards. The scenario you just described, where a company was reported by people to have had exceeded the standard but is found to be compliant by inspectors, has two possible explanations. First, the discharge of pollution fluctuates even during a company's normal operation, so compliance with the standard during inspection does not necessarily mean compliance the whole time. Second, some companies that know about the inspection beforehand may deliberately manipulate their production and pollution to temporarily deceive the inspectors into believing that they meet the standards. Targeting the second scenario, we take various approaches, including the satellite, drone, and automatic monitoring equipment installed in some enterprises more prone to noncompliance, to avoid being deceived by them deliberately. Moreover, people can call the tip-line 12369 and report pollution as soon as possible, then inspectors may rush to the scene and catch the polluters red-handed.

On top of that, due to limited monitoring capacity, local inspectors may miss some particular pollutants (nonconventional pollutants) harmful to health. We have required local inspectors to monitor and analyze all the indicators specified in the pollution standards. However, this will take time as local monitoring capacity comes short.

Regarding your second question, pollution permit is the only legal certificate that allows a stationary source to release pollution. It is a basic mechanism. Explicit requirements on pollution emission or discharge and rules to be followed are made to enterprises before the permit is granted. To align this mechanism with environmental standards, measures will be taken. First, some management standards necessary to the permit are not put in place yet as this is a new mechanism, and we will add them as soon as possible. Second, likewise, we will add the essential technical standards as soon as possible to ensure that future environmental standards and the permit can be aligned seamlessly.

Reuters: How much will China spend in the next five years to meet the environmental standards set for the 13th Five-Year Plan period? The first-phase implementation of the *Action Plan on Air* is coming to an end. Has the Chinese government considered a similar plan for the next 4–5 years? Is the Chinese government confident that the emission or discharge of all pollutants will meet environmental standards? What is the future plan and funding for that?

Zou Shoumin: Our annual budget for the formulation of environmental standards is estimated at around 30–50 million yuan during the 13th Five-Year Plan period. However, it is hard to put a figure on the total social cost if all the standards are met. We are striving to introduce about 800 environmental standards in the next five years, as proposed in the specific plan. The purpose is to fill in some gaps and update the emission and discharge standards that are more than five years old. The focus is to toughen air pollutant emission standards to improve air quality.

Regarding your second question, this year is the final year for the first-phase implementation of the *Action Plan on Air*. Relevant departments are evaluating the implementation and will present their conclusion. As for the rolling wave planning, the State Council has made its reply to the 13th Five-Year Plan for ecological and environmental protection, which sets forth major air-related goals and tasks by 2020. There will be a new plan following the *Action Plan on Air* seamlessly.

Wu Xuefang, Director of the Institute of Environmental Standards, Chinese Research Academy of Environmental Sciences

Regarding your third question, MEP has issued *the Notice of Implementing the Plan for All Industrial Pollution Sources to Meet Standards*, according to which all kinds of industrial pollution sources would gradually meet the standards over the 5-year span by the end of 2020. It would be a process during which enterprises would pursue technological upgrading and emission standards and corporate investment would change for the better. You can see that fewer enterprises are deliberately exceeding the standards, which lays the foundation for all pollution sources to meet the standards by 2020.

Guangming Daily: Where are we with the formulation of soil environment quality standard for the 13th Five-Year Plan Period? And what difficulties have we encountered?

Wu Xuefang: As Director General Zou has mentioned, the existing soil quality standard was released in 1995. Later, the urgently needed standards for environmental management of contaminated sites were put in place as a supporting standard. As the soil quality standard is very old, MEP has been working on its amendments for several years, and organized 20 plus meetings and workshops to discuss soil standards and their future implementation. Up till now, it has drafted two standards, one is on agricultural land and the other construction land.

Until now, three rounds of opinion canvassing have been carried out. The amendment is difficult and intense efforts are being made to refine it. According to the 13th Five-Year Plan, we will release and implement the amended version this year. Specifically speaking, the difficulty in the drafting process lies in the fact that standards of soil pollution, unlike those of water and air pollution, are formulated based on risk management principles.

CCTV: This year's government work report proposed the target of replacing coal with electricity and gas in three million households in northern China. Where are we with this? Will different environmental standards be formulated for the development of Xiong'an New Area?

Zou Shoumin: Your first question should be directed to the Air Pollution Control Department. So far as I know, the projects are being carried out on a tight schedule. There is a quota for Beijing, Hebei, and Tianjin, respectively, which will be assessed at the end of the year. I believe that these local governments pay great attention to air pollution, particularly heavy pollution in winter. As you can learn from news and other media coverage, we hold admonitory talks with underperforming governments. I believe it should not be difficult to meet the target of replacing coal with electricity and gas in three million households. As for the specific progress, you can interview the Air Pollution Control Department.

Regarding your second question, Xiong'an New Area is a major development strategy of the central authorities and an important arrangement for the next decade. The need to introduce tougher environmental standards for Xiong'an has not been considered for now. However, it is worth pointing out that Baoding, of which Xiong'an is a part, is itself a heavily polluted area and is enforcing special emission limits of air pollutants. Also, Hebei goes out of its way to protect Baiyangdian, a part of Xiong'an New Area, by strict water pollutant discharge standards, but we're not considering a special standard for Xiong'an New Area.

Legal Daily: We have so many national environmental standards, how well are they being observed? Is there any difference between the national standards and control standards? Can the standards available meet the regulatory requirements? Are there any gaps?

Zou Shoumin: Regarding the implementation of the environmental standards, two years ago we started to evaluate the implementation of emission standards for key pollutants, especially old standards. For example, we have evaluated the over-5-year-old standards of ten industries such as paper-making and thermal power. Overall, the implementation situation is acceptable to us. The regulatory departments have been strictly enforcing the standards in their on-site inspection, environmental assessment, and management. However, some companies complain that certain standards are too strict to be implemented in full. It takes time for the standards to be met by all. When drafting the standards, we have factored into human health and technological upgrading concerns. It is impossible that the standards are so lenient that all enterprises can meet them without much effort. Some companies have to take pains.

The percentage of enterprises meeting the standard is on the rise. The Bureau of Environmental Supervision and Inspection offers specifics and data, so I would not elaborate on the issue. Overall, the degree of leniency and the pollutants covered by the standards are consistent with China's national conditions and development phase. If enterprises make an effort, they can meet the emission targets.

Regarding the difference between national standards and control standards, what we formulated for most industries is national discharge/emission standards. The standards for solid waste layout not only pollutant limits but also many technical specifications and operational requirements. That is why they are called pollutant control standards, for example, *Control Standards for Pollutants from Domestic Waste Incineration*. In essence, national standards and control standards are the same.

Indeed, certain gaps exist in environmental standards, and there are mainly three reasons. First, scientific progress prompts people to have a higher expectation for the government, and they want us to control some newly emerging pollutants while some existing standards are too outdated, which leads to some gaps in management. Second, some standards, having been implemented for a while, are no longer sufficient for improving environmental quality and need to become more stringent. Third, technological progress weakens the relevancy of some standards and we need to revise them regularly. During the 13th Five-Year Plan period, we will strive to catch up on key standards, particularly supporting standards on monitoring methods.

Health News: Last year's National Sanitation and Health Conference set forth the strategic goal of Healthy China. My question is, of the environmental standards proposed in the 13th Five-Year Plan, which are directly related to health?

Wu Xuefang: The first article of *Environmental Protection Law* specifies that the goal of China's environmental protection is to improve the environment, prevent and control pollution and protect people's health. When we formulated standards, be it standard of air quality, surface water or soil quality, the starting point, also an important consideration, is health. Another important factor is environment.

When we set the emission limits of toxic pollutants, the ultimate goal is also health. As cars become more common, they are increasingly interwoven with public life. As early as in 2011, MEP issued the assessment guideline for passenger vehicle interior air quality. Several years ago, MEP started to revise the standard in this regard and has canvassed opinion once. First, the limit of some pollutants very harmful to health, specifically speaking, toluene, xylene, and ethylbenzene, will be adjusted. Second, we will toughen the pre-shipment consistency inspection of vehicles. Third, we will strive to make the currently recommendatory standards mandatory, which according to the 13th Five-Year Plan will be issued and implemented this year.

China News Service (CNS): The amended ambient air quality standard added $PM_{2.5}$ and ozone indicators, which helps China's emission standard keep up with the standard of an international organization, but which international organization is it? Are China's limits of pollutant indicators consistent with this standard?

Zou Shoumin: The international organization is the World Health Organization (WHO). The existing air quality standard in China is on a par with a more lenient version of the Interim Target-1 (IT-1) in the WHO air quality guidelines. It is at about the same level as India and Mexico but below that of the EU and USA's standard should be based on its environmental quality and economic development. Many Chinese cities cannot even meet the IT-1 standard, especially in heavy air pollution situations. Otherwise, heavy air pollution will not be so frequent in winter.

Shanghai Securities News: A key task for the 13th Five-Year Plan period is to issue the *Evaluation Guideline for the Implementation of National Pollutant*

Emission Standard. Is there a timetable for this? And what are highlighted in the guideline?

Zou Shoumin: In September 2016, we issued the guideline. You may download the text on MEP's website.

Wu Xuefang: The standard implementation evaluation is an integral part of environmental standard work. The *Law on the Prevention and Control of Atmospheric Pollution* and other laws provide for such regular effectiveness evaluation on environmental quality standards and emission/discharge standards. We evaluate the implementation effects, the scope of application, selection of pollutants, scientificity of the limits, and feasibility of standards. We also factor in problems encountered during actual law enforcement. Overall, based on this systemic evaluation, the guideline comes up with specific coping methods and requirements. Also, on the basis of the evaluation conclusion, we determine what the emission standards should be and whether we need to revise quality standards.

People's Daily: What are your considerations when you said the amendment of environmental standards is health centered? The drafting or amendment of some standards specified in the 12th Five-Year Plan has not been completed, why is that? Will a prospective approach be adopted for standard formulation and amendment in the 13th Five-Year Plan period?

Zou Shoumin: The *Environment Protection Law* explicitly stipulates that the purpose and ultimate goal of environmental protection is to protect people's health. When we formulate standards, particularly environmental quality standards and emission standards, we always consider their influence on health. For example, the pollutants and limits in the water quality standards are decided based on environmental criteria as well as other national and international standards. We consider the influence of standards on both health and the environment. In the case of water standards, we consider both health and certain aquatic organism. This is made very clear by the standards we set especially the indicators for certain aquatic organism.

Regarding your second question, there are some reasons why some standard formulation or amendment tasks set for the 12th Five-Year Plan were not completed.

First, we were not forward-looking enough when formulating the standards. It often takes a long period to complete making a standard, during which social and economic progress and new environmental management need to place new requirements on it. We did not foresee the new changes and had to make an adjustment, which prolonged the formulation cycle.

Second, as standards are being implemented, technology can change greatly and rapidly. As new technologies are deployed, new indicators will be made and the relevant standards need to be updated.

Third, the agencies charged with formulating standards may undergo personnel turnover or simply do not value the task enough and thus fall behind schedule. In response, we will take targeted measures and strengthen project management. The unfinished projects will be completed and urgently needed new standards will be formulated or amended to put in place a complete standard system that can provide technical support for environmental departments to enforce the law and for enterprises to comply with the law.

The Beijing News: You just said that the purpose of formulating standards is to protect public health. The public are very concerned about standards on indoor air quality standards and air cleaners. Can you tell us something about them? The Environmental Development Center under MEP issued a technical standard for environmental labeling of air cleaners, which took effect on January 1 this year. Is health concern addressed in making this standard?

Zou Shoumin: Indoor air quality standard is indeed a national standard, but its formulation was led by the National Health and Family Planning Commission (NHFPC). MEP contributed to this process. Therefore, the amendment will be conducted by NHFPC in collaboration with relevant departments. I am not aware of their plan on this. An air cleaner is a product, thus its quality standard should be formulated by quality inspection departments. The *Technical Requirement for Environmental Labeling Products-Air Cleaner* issued by MEP is a standard for better-performing and more environmental-friendly air cleaners. The standard is referential for the public and available on MEP's website.

Liu Youbin (Moderator): Thank you for your support. The World Environment Day is approaching, and this year's theme is "Lucid Waters and Lush Mountains Are Invaluable Assets". The World Environment Day is a statuary day to celebrate the environment and its purpose is to disseminate information of environmental protection, encourage the public to engage and raise their awareness. We sincerely hope that the media friends can take this opportunity to disseminate scientific and legal information on environmental protection and help us fulfil that purpose. That is all for this regular press conference. Thank you!

Regular Press Conference (June)

June 21, 2017

On the morning of June 21, the MEP held the June press conference, introducing the soil environment management progress from 2016. Qiu Qiwen, Director General of the Department of Soil Environment Management of the MEP, together with Lin Yusuo, Director of the Research Center for Soil Pollution Prevention and Control, Nanjing Institute of Environmental Sciences, attended the press conference to share related information and answer questions from the media.

Ministry of Environmental Protection of the People's Republic of China,
2017 Press Conference Records of Ministry of Environmental Protection, the People's Republic of China, https://doi.org/10.1007/978-981-13-7330-5_7

MEP June Press Conference (1)

MEP June Press Conference (2)

Liu Youbin (moderator): Dear friends from the press, good morning! Welcome everyone to the MEP June Press Conference.

The soil is a physical foundation of the sustainable development of economy and society. Soil environmental quality is closely related to public health and the building of a beautiful China. Soil pollution prevention and control is an essential part of ecological civilization and environmental protection. In May 2016, the State Council printed and distributed the *Action Plan on Soil Pollution Prevention and Control*, or the Action Plan on Soil for short. It provided instruction on soil pollution prevention and control in China for the current stage and years ahead and received great attention from the public.

For today's press conference, we invited Qiu Qiwen, Director General of the Department of Soil Environment Management Department of the MEP, and Lin Yusuo, Director of the Research Center for Soil Pollution Prevention and Control of the Nanjing Institute of Environmental Sciences, to share with you relevant information about the implementation of the action plan and answer questions from reporters.

Let's welcome Director General Qiu Qiwen to share with us some information.

Qiu Qiwen: Good morning, everyone! Thank you for your consistent concern and support for the environmental protection effort in China. Soil environmental quality is related to the sustainable development of economy and society. It also has a great influence on people's health. It is an important part of promoting ecological civilization and safeguarding national ecological security. The Party Central Committee and the State Council attached great importance to soil pollution prevention and control. On the last May 28, the State Council printed and distributed the *Action Plan on Soil Pollution Prevention and Control* (the Action Plan on Soil), which will be the action guidance for the soil pollution prevention and control work in China for the current stage and years ahead. For 1 year, the MEP mainly implemented the following measures: first, established a working mechanism. We took the lead to establish a National Inter-Departmental Coordination Group on Soil Pollution Prevention and Control, developed work regulations and recent priorities, printed them out and distributed to institutes for implementation. Related departments established implementation plans, respectively, for their key work.

Second, fully kicked off the detailed survey of soil pollution. We carried out the *General Plan of the Detailed survey of National Soil Pollution* together with the Ministry of Finance, the Ministry of Land and Resources, the Ministry of Agriculture, and the Commission of Health and Family Planning, printed it out and distributed to local governments for implementation. Pushed forward the establishment of technical documentation of the detailed survey, lab selection, and personnel training etc.

Third, promoted the regulation standard system. Actively cooperated with the NPC to promote the legislation work of the *Soil Pollution Prevention and Control Law*. Issued the *Administrative Measures for the Soil Environment of the Contaminated Land Parcel (Trial)*, which will come into effect on July 1. Established the *Administrative Measures on the Soil Environment of Agricultural Land (Trial)* together with the Ministry of Agriculture. We have collected opinions from various departments, it will be implemented as soon as possible after getting the approval according to procedures. Revised and established soil environment standards of lands for agriculture and construction use, publicly solicited opinions 3 times. It is planned to be implemented before the end of this year.

Qiu Qiwen, Director General of the Soil Management Department of the MEP

Four, carried out the pilot demonstration of soil pollution prevention and control. Based on the 14 pilot projects, instructed local departments to launch a new batch of pilot projects for the application of soil pollution treatment and remediation technologies. Guided local governments to carry out the construction plan for 6 pilot areas of comprehensive soil pollution prevention and control.

Five, established the soil environment monitoring network. Carried out the general plan of soil environment monitoring and the deployment of nationally controlled monitoring points, clarified the layout of about 20,000 basis points, covering 99% of counties, 98% of soil types, and 88% of major grain producing areas, established the soil environment monitoring network for the initial phase.

Six, enhanced the achievement appraisal of the Action Plan on Soil. Organized to compile the breakdown plan of major goals required by the action plan, pushed different provinces (districts, cities) to sign the letter of commitment.

For the next step, we will mainly focus on the following tasks to comprehensively implement the *Action Plan on Soil*: first, fully conduct a detailed survey of soil pollution status to get to the bottom of the issue. Second, continuously support the legislation of soil pollution prevention and control, expedite standard establishment and revision. Third, establish the technology system for management, strengthen the classification management of agricultural land and the access management of construction land. Fourth, solidly implement pilot projects of soil pollution treatment and remediation and construct pilot areas for comprehensive prevention and control of soil pollution. Fifth, carry out the appraisal measures of the *Action Plan on Soil* to strengthen and supervise the assessment.

From practices domestically and abroad, soil pollution was not generated overnight, neither can it be solved at one stroke. We not only need to be prepared for a tough fight, but also have the patience for a long haul. We are still at the initial stage of soil pollution prevention and control in China, which means we need to set a solid foundation, pin down the priorities, focus on the baseline, and implement measures steadily. We hope and believe that we can reach the targets of national soil pollution prevention and control as planned through the support from the public with our joint effort.

Thank you very much!

Liu Youbin (moderator): Friends from the press, you may ask questions now.

National Business Daily: It is said on the internet that the *Action Plan on Soil* may generate a market with the value of more than 500 billion RMB for soil remediation, which attracted more and more companies to join this industry. What measures will the country adopt to drive the development of this soil treatment and remediation industry?

Qiu Qiwen: Thanks for your question. On the last May 28, the State Council printed and distributed the *Action Plan on Soil*, stating that soil pollution prevention and control should focus on prevention and prioritize protection while considering risk management. Aggressive management and remediation are not encouraged. Instead, the priority should be orderly treating and remedying the contaminated lands that are planned to be used for constructing residential communities, commercial areas, schools, hospitals, nursing homes, etc. This solution has taken the decades of soil treatment and remediation experiences and lessons learned abroad as references.

Currently, the domestic soil remediation industry has just started. We need to establish a complete industrial chain from environmental investigation, risk evaluation, risk management, to soil pollution treatment and remediation, and the assessment of remediation effects. It is important to promote several leading companies with strong comprehensive capabilities and a batch of active SMEs to boost the soil pollution treatment and remediation effort.

As the market just started to develop, the central government will strengthen the supervision from the following aspects. First, information disclosure. The MEP issued the *Administrative Measures for the Soil Environment of the Contaminated Land Parcel (Trial)*, requiring that all major contents of investigation reports, risk assessment reports, risk control plans, project plans of treatment and remediation, assessment reports of the effect of treatment and remediation need to be disclosed on the internet, available for public supervision. Second, strengthen credit constraints. Standardize the management of companies and personnel working on soil pollution treatment and remediation and establish a comprehensive supervision mechanism to publicly disclose the list of companies with a lack of technology service capability, operation and management competence, and comprehensive credit performance, through the company credit information publicity system. Third, emphasize accountability. The MEP is working hard on establishing a lifelong accountability measure for soil pollution treatment and remediation, which is planned to be launched by the end of this year. Thank you.

Southern Metropolis Daily: Statistics from the public security department showed that cases of dumping hazardous wastes across borders have been increasing every year. There have been more cross-regional cases showing the trend of industrialization. Why are there still so many cases of illegally dumping hazardous wastes despite repeated prohibitions? How to effectively crack down on the issues of dumping wastes at another place?

Qiu Qiwen: Thank you. The environment management of hazardous wastes started in the 1990s in China through implementing the *Basel Convention*. The regulation system for hazardous wastes treatment we have today is built on more than 2 decades of practices and the great experiences and measures of European and American countries. During the recent years, the MEP mainly promoted the environment management of hazardous wastes from the following aspects. First, promoted the environment management system of hazardous wastes. We issued the *Directory of National Hazardous Wastes* together with other related departments, established the exemption management system, and supported the revision of the *Law on the Prevention and Control of Environmental Pollution by Solid Wastes* by the NPC. Second, built up the capability of hazardous wastes treatment at various places. So far, the environmental protection system granted 2,034 business permits of hazardous wastes treatment nationwide. Certified companies now treat a total of 52.63 million tons of hazardous wastes annually, increased by 29.38 million compared with 2010. Third, strengthened the standardized management appraisal of hazardous wastes. During the "12th 5-year Plan", we selectively inspected more than 8,000 companies related to or generated hazardous wastes, the pass rate increased by 15%. Fourth, promoted law enforcement effort. In 2016, we launched the special campaign against illegal acts related to hazardous wastes together with the public security department, examined 46,397 companies related to hazardous wastes, 1,539 cases were filed and investigated, and 330 cases were transferred to the public security department to investigate for criminal responsibility.

The major reason for the frequent occurrence of illegally transferring and discharging hazardous wastes in China is that some companies are reckless with greed, so they decided to take the risk, secretly discharge wastes, and illegally transfer hazardous wastes. It shows companies did not have enough awareness of law-abiding and failed to take on their due responsibilities of environmental protection.

Next, the MEP will strengthen the supervision of hazardous wastes issues from the following aspects. First, speed up the establishment of the management system, promptly push forward the revision of regulations including the *Law on the Prevention and Control of Environmental Pollution by Solid Wastes*, and clarify the responsibilities of the companies generating hazardous wastes to harmlessly dispose of the wastes and their lifelong penalty. Second, clarify companies' responsibilities as legal entities and government's supervision liability, establish a publicity platform for companies' environmental protection information to expose and strictly penalize the companies with illegal acts. Involve the management situation of hazardous wastes as a part of educational supervision and lifelong environmental protection liability, enhance the accountability of leadership of the party and government. Third, enhance law enforcement and establish a long-term mechanism to crack down on the illegal acts related to hazardous wastes. We need to strengthen the connection between administrative law enforcement and criminal justice and maintain the intensified attitude of zero tolerance to illegal acts related to hazardous wastes, strictly crack down on such acts through the joint effort with the public security department. Thank you.

Financial Times: As far as I know, some local governments are reluctant to disclose the situation of local soil pollution, because if they disclose the information, people may no longer purchase local products. I am wondering to what extent will you disclose the related details of such soil pollution? Thank you.

Qiu Qiwen: Thank you for the question. In my opinion, all countries, including China, pay great attention to the issue of soil pollution information disclosure. We are organizing the detailed survey of national soil pollution right now, which is an essential task of the *Action Plan on Soil*. The major requirement is to clearly understand the size and location of the pollution of agricultural lands and its impact on the quality of agricultural products, as well as the location of the pollution of commercial lands in key industries and its related environmental risks. According to the *Action Plan on Soil*, by 2020 when the detailed survey of national soil pollution is finished, we will disclose the result to the public following some procedures. As to your question about to what extent will we disclose the information, I think we will follow the established practices of all countries to disclose the information based on the inspection result. Generally speaking, we will disclose information as much as possible and adopt effective measures for management and control. Thank you.

China Radio International: I noticed that the newly issued announcement by the MEP mentioned situations related to air and water pollution, but soil pollution was not covered. Could you please talk about the size and management progress of soil pollution in China?

Qiu Qiwen: Thanks for the question. In 2014, the MEP and the Ministry of Land and Resources jointly issued the *Report on the Inspection of Soil Pollution*. The inspection result showed that the overall status of soil pollution in China is not so optimistic. In some places, soil pollution is very severe and the environmental quality of farmland is quite worrying, while soil environmental issues of industrial and mining wastelands are significant. Generally, 16.1% of the land in China exceeded the pollution standard according to the inspection of sampling points, in which the percentage of minor, light, moderate, and severe pollution are respectively 11.2%, 2.3%, 1.5%, and 1.1%. I would like to mention that the result is based on the inspection of sampling points. Since the soil quality may vary a lot from points to points, this result does not represent the percentage of contaminated land. It just shows the overall soil pollution status in China from a macro perspective. Currently, we are cooperating with relevant departments to implement the national detailed survey of soil pollution, further clarify the size and location of soil pollution and set a solid foundation for soil pollution prevention and control.

In terms of this year's environmental status announcement you just mentioned, there was relatively little information on soil environmental quality covered. The collection and disclosure of soil environmental data depend on the construction progress of the national soil environmental quality supervision network. Now the network in China is still under construction and the monitoring and supervision work is still at the initial and pilot stage. I mentioned earlier that we have deployed more than 20,000 sampling points. By the end of 2020, the sampling points will cover all counties (cities and districts). In my opinion, through promoting the construction of the national soil environmental quality supervision network, we will be able to disclose soil quality information through the environmental quality announcement very soon. Thank you.

Reuters: What are the supporting fiscal and political measures for soil pollution prevention and control? Could you please share with us some successful cases about soil pollution treatment, especially through fiscal support? How much time and money have been spent on the remediation?

Qiu Qiwen: Thank you. The Chinese government attached great importance to soil pollution prevention and control. In terms of financial support, the finance department of the central government established the fund dedicated to soil pollution prevention and control. From the last year to the first half of this year, the central government provided about 14.6 billion of fund to support the local soil pollution prevention and control effort. Related departments like the Ministry of Science and Technology established plans to provide technological support, planning to prioritize and kick off the key project of soil pollution prevention and control. Frankly speaking, related departments are promoting and implementing relevant policies from the financial and technology perspectives to support the work on soil pollution prevention and control.

You asked another question regarding the cost of successful cases, measures adopted, and their effects. I believe Mr. Lin Yusuo can answer this question from a more professional perspective.

Lin Yusuo, Director of the Research Center for Soil Pollution Prevention and Control of the Nanjing Institute of Environmental Sciences

Lin Yusuo: Good morning. Everyone knows that soil treatment and remediation is a tough issue for the whole world. Foreign countries started this task earlier than China did. Hence, when we are working on soil pollution treatment and remediation in China today, on the one hand, we can leverage the successful experiences from abroad and learn from their good practices; on the other hand, we are utilizing the successful experiences and cases of soil improvement and pollution prevention and control domestically.

You asked if there have been any successful cases in China, I would like to give you a brief introduction. It is fair to say that China has great experience and cases in terms of agricultural soil pollution treatment. Here is an example: in the 1970s and 1980s, farmlands were contaminated by BHC. In order to solve this issue, the Chinese government banned the use of BHC in 1983 to eliminate pollution from the source. By adopting this measure, we followed the law of nature that organic pollutants are bio-degradable, strengthened supervision and the effort of soil improvement. More than 3 decades have passed, we kept monitoring the level of BHC residual in farmlands every 10 years, and we found out that the level has returned to the safety range. This is a successful case related to farmland contamination risk control in China, which was also recognized globally.

In addition, in the past 2 decades, China actively explored remediation technologies for heavy metal contamination. China is now at the leading position internationally regarding the research level and technology application on big-scale farmlands. We believe China is doing quite well in this aspect. For the remediation of polluted lands, during the past decade, remediation technologies applicable to contaminated soil and underground water developed quickly with great achievement. They have been successfully applied to many big-scale polluted areas with complex issues in China. Take chemical oxidation technology as an example, it enjoyed significant development in China in terms of the application of this technology, the R&D of related equipment, and the establishment of a professional service system. That is all for now, thank you.

The Paper: My first question is that you mentioned the construction of a soil pollution monitoring network, how can the deployment of those monitoring points coordinate with the current ones deployed by the Ministry of Agriculture and the Ministry of Land and Resources? The second question is related to the heavy metal contamination in the "Manganese Delta" at the junction of Guizhou, Hunan, and Chongqing.

On April 12, 2017, when the central government provided Chongqing with the feedback from its environmental inspection, it mentioned that there were 18 companies at Xiushan, Chongqing, in the electrolytic manganese industry, that lacked anti-seepage processes. When we were investigating in the "Manganese Delta" area in April, we found out similar issues in Songtao, Guizhou, and the heavy metal residual, including manganese, in the sampled water and soil at the downstream area of these companies significantly exceeded standards. According to the requirements of the *Action Plan on Soil*, it is required to investigate soil pollution and clearly understand the quality status. Which industries have been covered in the detailed pollution inspection by the MEP? What measures will be implemented for the next step to solve the pollution issues discovered in companies and regions?

Qiu Qiwen: This is a rather long question. I will try to cover all points. The detailed survey of national soil pollution status mainly focused on key industries related to soil contamination, including nonferrous mining, refining, oil exploration, oil processing, chemical industry, pesticide, coking, electroplating, and leather production. The electrolytic manganese industry was also involved. Recently, local environmental protection departments at the "Manganese Delta" area led the effort and completed the initial deployment of soil pollution inspection and identified key polluting sources. The "Manganese Delta" area is the key place stated in the *"12th Five-year Plan" of the Comprehensive Prevention and Control of Heavy Metal Pollution*. The MEP has guided the implementation of related pollution treatment effort in the electrolytic manganese industry according to the plan and achieved initial results. For the next step, the MEP will push forward the establishment of a comprehensive prevention and control coordination mechanism in the "Manganese Delta" area. In addition, we will also urge local governments to strengthen the rectification against "the companies that are in unauthorized locations, lacked the proper certificates and failed to meet emissions standards," and the treatment of pollutant residual in manganese production sites and other plants. Thirdly, we will emphasize companies' responsibilities as legal entities, strengthen law enforcement and supervision, to ensure that the electrolytic manganese industry can meet the emission standard of pollutant sources. At the same time, we will take the result of a detailed survey of soil pollution into consideration, use it while conducting further inspections. Especially for the agricultural land with a high risk of soil pollution, we will ask local government to study and establish work plans and timely adopt risk control measures to ensure the safety of farmlands and construction lands. For the company lands with a high risk of soil pollution, we will provide targeted pollution prevention, control, and risk management measures while considering the detailed survey result of these lands.

You also mentioned the issue of coordination with other departments in terms of deploying monitoring points. I covered this earlier, the MEP is leading the effort to carry out the general plan of the national monitoring system of soil pollution and deploy the monitoring points according to the requirement raised by the State Council. Now, the MEP is cooperating with other related departments including the Ministry of Agriculture and the Ministry of Land and Resources, following the principle of integrated planning and standard, as well as unified information collection and disclosure processes, to have these 2 ministries screen and select the monitoring points under regular national control. It means that these 2 ministries will select fundamental monitoring points that are organically connected to the soil environment monitoring network established by the MEP while sharing all information.

Jiemian.com: Director Qiu, based on the inspection result currently available, what is the estimated cost for the remediation of farmlands in China? According to the regulation and related document, soil pollution remediation mainly follows the principle of "the polluters are responsible for the treatment", and "the developers are responsible for the management", but what if the polluter does not want to shoulder the responsibility and pay for the cost?

Qiu Qiwen: Thanks for the question. In my opinion, compared with the treatment and remediation of polluted lands, the measures for farmlands can be different. Because it is required that after the treatment and remediation, farmlands should still be able for agricultural use, which means there are some limits to the remediation technologies applicable to farmlands. The cost of farmland pollution treatment and remediation can vary dramatically according to the different types of pollutant, the significance of pollution and the technologies adopted. To answer your question about cost, based on the estimation from experts and past experience, the cost can be a couple of thousand to dozens of thousand every mu (the unit of area that is often used in South Asia. 1 mu corresponding to 1/15 ha).

According to the principle of "the polluters are responsible for the treatment," polluters should take the responsibility for treatment and remediation as the main legal entity. In the case of loss of the responsible subject or it is unclear, the local people's government at the county level will need to shoulder related responsibilities based on regulations.

The treatmcnt and remediation of soil pollution are always difficult, requiring a large investment for a rather long period of time. The goals of such an effort should be defined based on the purpose of use of the land. For the land that is going to be developed as residential areas or public facilities such as hospitals, schools, nursing houses, etc., environmental investigation, risk evaluation, treatment, and remediation are required to meet the soil environmental quality standard before any development and usage. In terms of the lands that are temporarily not for development or lack the condition to do so, risk control and management will be the major focus. For example, we can set up risk control areas for environmental supervision and monitoring. If any spread of pollution is discovered, measures like blocking or isolation should be taken to control the risk. So in my opinion, the focus of soil pollution prevention and control is risk management. Thank you.

New Beijing Daily: You mentioned soil pollution treatment is different from the treatment of air or water and no big-scale treatment or remediation will be applied. Then how to control the risk of the existing soil pollution? How to treat and control the lands that are not polluted or are being polluted? You also talked about pollution issues related to BHC, which can be brought to the safety level through natural degradation. What should we do with the polluted soil? Will they be used in the future? My second question is about the draft of the *Soil Pollution Prevention and Treatment Law* which is being reviewed by the Standing Committee of the NPC. Could you share with us some highlights of this draft? As for the requirement that when the responsible subject cannot be defined, the local people's government should shoulder the responsibility, will it be stated in the *Soil Pollution Prevention and Treatment Law*? In addition, will there be clear regulations prohibiting development in the areas where risk management and control measures are implemented?

Qiu Qiwen: OK, thanks. You asked 3 questions and I will answer the first and the third one, Director Lin will answer the second one. The first question is about risk control. The focus of soil prevention and treatment is risk control, which is different from the treatment of air and water pollution. But what is a risk? It includes 3 elements, one is the pollutant, the source; another one is the channel, or say the pathway of exposure; the third one is the recipient, meaning the object under protection. The objects under protection are people and crops. For soil pollution, in particular, cleaning up the source is the correct approach for treatment and remediation. But it is still not good enough, we should also focus on how to cut the channel of exposure, and how to establish reasonable strategies based on the goals of protection. The logic of cutting the exposure pathway is just the same as using sunscreen to block the sunlight. As long as the exposure pathway is blocked, people will not be able to contact polluted soil, hence the risk is under control.

We can also adopt measures of agronomy regulation, substitute plantation, structural adjustment of planting, restoring farmland to forests or grasslands, to use the land safely and effectively and control risks, ensuring the safety of agricultural products. In terms of the substitute plantation, for example, farmers can plant corn if rice cannot grow, or even switch to plants like cotton if crops cannot grow. Blocking and isolation measures can be adopted for the construction lands. We visited a waste storage site in Germany, which was covered by geomembrane to prevent rainwater seepage; at the same time, they extracted underground water to lower the water level and kept the water from contacting wastes to prevent pollution. That is how to control the risk of underground water pollution. In addition, we will adopt measures to monitor the underground water around the contaminated land. Once any spread of pollution is discovered through underground water monitoring, we will implement countermeasures immediately. Therefore, in my opinion, risk control means to adopt different measures according to different purposes of lands. The core is to control soil pollution risks and realize the safe usage of lands, instead of solely relying on enormous capital investment for excessive treatment and remediation. On the other hand, we will fully remedy the lands if needed. As I emphasized just now, weak remediation is not acceptable either, the core is to ensure safety. Director Lin will answer the question about BHC.

Your third question was about the *Soil Pollution Prevention and Treatment Law*. The Standing Committee of the NPC has attached great importance to this law from when the project was just initiated and categorized it as a level 1 legislative item. It took great effort and received great achievements. The law will be submitted to the NPC for group deliberation in short. We learned that related departments of the NPC will hold a specific briefing to release more authoritative information. You can stay tuned for more information.

Lin Yusuo: I will further elaborate on these issues. Director Qiu talked about risk control, which does not mean that we are going to leave the lands as they currently are. Scientific risk management and control should be based on the knowledge of the characteristics of pollution and its pattern of change. Take the case of BHC as an example. The reason why we adopted risk management and control measures which allowed the farmland to be utilized continuously to grow crops after banning the use of BHC is that we have a relatively comprehensive understanding of BHC. Firstly, major pollution on crops caused by BHC is when it is directly sprayed on plants during its use. This is the main pathway of pollution. So after it is banned, it is guaranteed to a certain extent that crops will not be polluted by pesticides. This is one concern.

The second one is about the BHC residual. We know BHC as an organic pollutant. Even though it is relatively hard to be degraded, as it is used in farmland, microorganisms still can fully degrade it. Its impact on crops will gradually decrease as the concentration goes down.

In addition, BHC can become a combination form in soil, which is favorable for detoxification. The departments of agriculture and environmental protection inspect the BHC residual about every 10 years. During the last national soil pollution inspection, as covered in the communique, BHC is no longer the major soil pollutant, nor the cause of safety issues of agricultural products. Therefore, I would like to emphasize that the risk management and control we adopted is based on the scientific knowledge and the best choice we can make according to the reality.

China News Service: Can you talk about the related situation of the detailed national inspection of soil pollution? Compared to 2005, has the soil pollution in 2013 getting better or worse? What are the causes of pollution? Is it related to the overused chemicals or fertilizers led by the pursuit of food production?

Qiu Qiwen: Thanks for the question. A comprehensive understanding of soil pollution situation is the essential foundation for implementing soil pollution prevention and supervision. From 2005 to 2013, the MEP and the Ministry of Land and Resources jointly carried out the First Survey of Soil Pollution Nationwide, got a basic understanding of the macro situation of soil pollution. Currently, we are implementing detailed surveys based on the exist result from former surveys to get further information about the size, location of farmland pollution, and its impact on agricultural products' quality. We will also get to know the location of polluted land used by companies in key industries and its environmental risks. Overall, we can collect authoritative and unified soil environmental information with high accuracy to establish a national soil environmental information management platform, which will lay the scientific foundation for a comprehensive implementation of the soil pollution prevention and treatment plan. Through the joint effort from various departments, it has been running quite smoothly in general. First, the general plan of the detailed survey is finished and submitted to the State Council. After getting the approval, it has been printed and distributed to local organizations for implementation. 31 provinces (regions, municipalities) have drafted the plan for local implementation. Second, a series of technical standards of the detailed survey has been established. The detailed survey is a systematic project with high technical requirements. A series of technical files is necessary to be carried out to serve as guidelines for the inspection. Up to now, we have officially published 1 technical file, 4 are fully established and waiting to be released, 8 are under revision, and another 8 are in compilation. Third, we completed the development and debugging of the information management platform and the information terminal for the detailed survey. Fourth, we developed the quality control system and led the effort in selecting inspection labs. We pinned down 37 labs for quality control and more than 200 for testing, which will be disclosed to the public recently. Fifth, we organized staff training. From the start of this year, nearly a thousand professional staff have received training from the MEP, together with the Ministry of Land and Resources and agricultural departments. Training is also carried out at the provincial level. Sixth, organized related pilot testing of technical standards. We conducted pilot testing of technical standards in Hebei including the Xiong'an New Area, Guangdong, Hunan, Chongqing, Guangxi among other provinces successively and set the foundation for the all-round detailed survey. Next, we are going to further implement the inspection according to the existing plan.

The detailed survey is different from the last inspection. The most significant feature is that it is conducted by five departments together, including environmental protection, finance, land and resources, agriculture, and the Health and Family Planning Commission. Especially through the close collaboration among the MEP, the Ministry of Land and Resources, and the Ministry of Agriculture, leveraging the inspection results from these 3 departments to avoid repetition, the accuracy and efficiency of our work have been greatly promoted. Second, we specified inspection objects. The first one was a general inspection that covered 6.3 million km^2 of land with a grid of investigation points. For this time, we are also inspecting soil pollution of the land used by companies from key industries instead of merely investigating farmland. Third, we upgraded inspection methods. We divided inspection objects into units and deployed investigation points in each unit based on pollution sources and its spread pathway to enhance the representativeness of the points. Four, we promoted inspection precision. During the first inspection, we deployed points based on a grid of 8,000 m * 8,000 m for farmland, and 32,000 m * 32,000 m for grassland. For the current one, farmland inspection point deployment was based on a grid of 500 m * 500 m, other lands are based on 1,000 m * 1000 m, significantly promoted its precision. Five, it's required that all surveys conducted must follow the same work plan, lab selection requirements, assessment standards, quality control, and duration.

China Youth Daily: You mentioned that the inspection on farmland will not have a result until 2018. However, some media covered the issue of wheat contamination in some provinces which is quite worrisome. Will inspections for emergency response available in the future?

Qiu Qiwen: Thanks for the question. In fact, reports about cadmium-contaminated wheat triggered public attention. We learned that the provincial government of Henan attached great importance to this issue and instructed the provincial agricultural department to kick off the food safety emergency response mechanism according to the *Measures for the Supervision of Grain Quality and Safety* issued by the NDRC in 2016, as well as the *Opinions of Pushing Forward the Implementation of the Safeguard Mechanism for Grain Quality and Safety* issued by the State Administration of Grain. Provincial agricultural department of Henan organized related experts to inspect the "cadmium-contaminated" wheat field on-site, sampled problematic wheat and soil and sent them to the agricultural product quality assessment centers (Nanjing, Wuhan, Zhengzhou) of the Ministry of Agriculture respectively for tests. The test results haven't been officially released. Local government in Xinxiang started to monitor the harvest, transportation, and delivery of the "cadmium-contaminated" wheat from end to end. All wheat is now securely stored after harvest, with installed monitoring devices and dedicated personnel for supervision. Further detailed information may be provided by relevant authorities later. You mentioned an issue about quickly identifying and taking countermeasures when a significant soil environmental issue is discovered. During this detailed survey, especially when we defined the investigation scope, we followed the principle of asking all local people's governments at city and county levels to include the areas with the most severe soil problem into the key areas of this detailed survey to fully understand the situation. Government at all levels will then adopt different measures of treatment and control to ensure the environmental safety of farmland.

China Daily: We noticed the heavy metal contamination in Hunan, is there any updates on the treatment? Changde City has been included in the pilot area of soil pollution prevention and treatment, is there any specific and scalable practices and experiences?

Qiu Qiwen: The State attached great importance to the issue of heavy metal pollution prevention and control you just mentioned. During the "12th 5-year Plan", the State issued the *"12th 5-year Plan" for the Comprehensive Prevention and Control of Heavy Metal Pollution*, which received significant achievements in general. I can share with you some numbers. By the end of 2015, the total emission volume of the key pollutants of five types of heavy metal (lead, mercury, cadmium, chromium, and metallic arsenic) decreased by about 1/4 than 2011, which is a great achievement. In terms of emergency accidents related to heavy metal, less than 3 incidents happened every year between 2012 and 2015 on average, significantly dropped from a dozen of cases in 2010 and 2011, which also reflected the effectiveness. Heavy metal pollution has its characteristics. It is caused by long-term accumulation. Just like the pollution in Hunan you talked about, it was also caused by years of accumulation. It will take time for its treatment and prevention, and eliminating its negative impact, it cannot be changed overnight. Next, the MEP will incorporate the implementation of the *Action Plan* on *Soil Pollution Prevention and Control* to continuously promote the effort of heavy metal pollution prevention and control and enhance the environmental quality in key areas, in order to address the environmental risks that concern the public.

As for if there is any good experiences or practices, in my opinion, the following experiences are available for the comprehensive treatment of heavy metal in key areas: first, clarify responsibilities. We need to uphold the principle of "implementing different policies for different areas, control new pollutions and treat the exist ones, the government provides guidance, and companies shoulder the responsibilities," and establish a responsibility system for environmental protection to solve the current problems while controlling new ones from happening. Second, all relevant departments will take joint effort for management and treatment aiming at enhancing environmental quality and focus on the targeted pollution treatment in key areas. Third, take the effort to adjust industrial structures, optimize industrial distribution, and upgrade technological levels. Four, create new investment and financing mechanisms to raise the fund for treatment through multiple channels. We can attract capital from the private sector through the innovation of investment mechanism while the government promotes its investment. In general, the State will continuously enhance its effort of heavy metal pollution prevention and control based on the progress made during the "12th Five-year Plan".

Science and Technology Daily: Agricultural planting, overused nitrogen fertilizer, and pesticide in soil will lead to pesticide residual in agricultural products and affect food safety for all. Is there any supervision mechanism to trace back the pollution source from the end? According to our survey, there is no soil testing applied in the rural area, including pesticide residual monitoring. Is there any relevant monitoring data available? We noticed that it is not very transparent in this respect. In the future, will there be a system to trace back the source of soil pollution through testing agricultural products?

Qiu Qiwen: Thank you. As I mentioned in the beginning, soil pollution prevention and control should "focus on prevention and prioritize protection while considering risk management". Soil pollution cannot just focus on the treatment at the end, controlling the source is more important. Why? After the soil is polluted, the treatment and remediation are extremely difficult and costly. Experiences overseas show a ratio of 1:10:100, which means pollution prevention may cost just 1 RMB, but risk control can cost 10 RMB, while the treatment from the end can cost 100 RMB. From this perspective, we need to firmly establish policies and measures prioritizing prevention. This is the most distinctive characteristic of soil pollution prevention and treatment. Because once the soil is contaminated, it is hard to be fully recovered. It is also why we have been emphasizing the importance of considering exposure pathways, protected objects, and defining different remediation goals for different purposes when implementing risk control.

You said that people are concerned about food safety. I am just the same as everyone else. Related departments of the State adopted numerous effective measures and a series of supervision approaches to solve this issue. For example, there are special supervision and treatment mechanisms for the grains with excessive residuals, from discovering issues, supervision, storage, to its treatment, everything is covered. But there is a dedicated responsible management department. I can just share some information that we know of for your reference.

China Business News: Recently, there was a pollution case of dumping of waste liquid happened in Wuji, Hebei. Did the MEP send specialists for on-site inspection? What is the latest result?

Qiu Qiwen: Thanks for the question. The Wuji accident was extremely serious in nature. We are very sad to hear about this issue. The MEP attached great importance to this case and promptly organized personnel from departments of environmental supervision and the soil protection for on-site instruction and urged the local government to conduct emergency response and inspection. Local government also took countermeasures immediately. Now the pollution is relatively under control with no spread. For the next step, we will analyze the test result and adopt targeted measures accordingly. Currently, the investigation is still undergoing, I believe we will disclose related details to the public in a timely manner when the inspection result is available.

Liu Youbin: That is the end of today's press conference. Thank you for your participation!

Regular Press Conference (July)

July 20, 2017

The MEP held a press conference on the morning of July 20. The topic for this conference is on international cooperation on environmental protection. Mr. Guo Jing, Director General of the Department of International Cooperation of MEP, Mr. Yu Lifeng, Deputy Director General of Foreign Economic Cooperation Office of MEP, and Ms. Zhang Jieqing, Deputy Director General of China-ASEAN Environmental Cooperation Center of MEP attended the conference, and took questions from the press.

Ministry of Environmental Protection of the People's Republic of China,
2017 Press Conference Records of Ministry of Environmental Protection, the People's Republic of China, https://doi.org/10.1007/978-981-13-7330-5_8

MEP held July Press Conference (1)

MEP held July Press Conference (2)

Liu Youbin (moderator): Good morning, friends from the press. Welcome to today's press conference.

The United Nations Conference on Human Environment held in 1972 charted a new course for modern environmental protection. It then became an international consensus that protecting the one and only planet earth is the shared responsibility of the mankind. The Chinese government also sent a delegation to this conference, and in the following year, convened the 1st national conference on environmental protection in Beijing. As an environmental power still at development stage, China not only commits itself to tackling domestic environmental issues, is also actively engaged in international environmental affairs to fulfill its due responsibilities of promoting global environment and development.

Today we are honored to have Mr. Guo Jing, Director General of the Department of International Cooperation of MEP, Mr. Yu Lifeng, Deputy Director General of Foreign Economic Cooperation Office of MEP, and Ms. Zhang Jieqing, Deputy Director General of China-ASEAN Environmental Cooperation Center of MEP with us. They will brief you on the latest progress in China's international cooperation on environment and development, and then take your questions. Now let's give the floor to Mr. Guo Jing first.

Guo Jing: 45 years ago in 1972, the United Nations had its first Conference on Human Environment in Stockholm, marking the first time in history when environmental issues were brought onto international cooperation agenda, and opening the door of international environmental cooperation. Nearly half a century on, the whole world is still struggling to balance the environment with development, and it's a daunting challenge to implement the 2030 Sustainable Development Agenda. In China, conflicts between environmental protection and development remain, and with mounting environmental pressure, degrading ecosystems and environment stood out as one of the major weak links in realizing a well-off society.

While dealing with domestic environmental issues, China highly valued and actively engaged in in-depth international cooperation on environmental protection, which is also one of the mandates of the MEP. To fulfill this role, the MEP has conducted cooperation and exchange with over 100 countries on environmental protection, signed nearly 150 cooperation documents with over 60 countries and international organizations, and established cooperation mechanisms with a number of countries, regional and international organizations, devoting itself to building a platform of collaboration. As such, a high-level, multichannel and broad-ranging cooperation and communication mechanism gradually took shape, playing a key role in protecting the environment, implementing international conventions, and helping other developing countries in doing better.

This is seen particularly in China's contribution to improving global environmental governance, promoting sustainable development, and implementing multilateral environmental agreements. The *Montreal Protocol on Substances that Depletes the Ozone Layer* has been universally considered the most successful multilateral environmental agreement so far approved and implemented by the international community. Under this framework, the amount of substance depleting Ozone layer phased out by China took up over 50% of the total of other developing countries combined, which was widely acclaimed by the international community.

In recent years, China has shifted its role from a learner, a participant, and a beneficiary in international environmental protection to one that is able to share experience, to promote and to actively contribute to global environmental protection. Over the past 5 years, China has put forward a wealth of new development concepts like ecological civilization and green development, and actively contributed China's wisdom and solutions to the world, which worked quite well and was highly valued by the international community. As is pointed out in the report of *Green is Gold, The Strategy and Actions of China's Ecological Civilization* released in 2016 by the UN Environment, "as China firmly supports and actively implements the concept and actions of sustainable development at the global level, its effort to build an Eco-civilization will make a significant contribution to the 2030 Agenda for Sustainable Development." The bold step taken by China in promoting ecological civilization will not only benefit itself, but will also be of valuable reference for those latecomers who are likely to be locked in the traditional development path, and will ultimately help to establish a new global environmental governance system.

China advocates, supports and actively participates in South–South Cooperation. As a large developing country, China stands ready and will do everything within its capability to support environmental protection within the framework of South–South Cooperation, and will continue to share its experience for the shared goal of sustainable development. China so far has introduced a "Green Envoys Programme", set up a China Trust Fund in the UN Environment for South–South Cooperation, developed the environmental cooperation strategy and action plan together with ASEAN countries, launched a ministerial meeting on China-Africa Environmental Cooperation, and conducted environmental policy exchanges with countries across South Asia, the Arab, Latin America, and South Pacific.

Sound environmental cooperation between China and the developed countries like the US, Japan, and Germany, etc., has enabled China to draw upon advanced ideas and experience and improve relevant technologies. It further boosted China's environmental protection industries, and positively contributes to China's overall ecological and environmental protection. China also contributed its bit by actively participating in subregional and regional environmental cooperation initiatives under the frameworks of tripartite meeting among China, Japan and ROK, BRICS, Shanghai Cooperation Organization (SCO), APEC, ASEAN plus China, Japan and ROK, north-west Pacific partnership, and Partnerships in Environmental Management for the Seas of East Asia (PEMSEA).

Mr. Guo Jing, Director General of the Department of International Cooperation of MEP

The Belt and Road Initiative (BRI) is important international public goods, of which ecological and environmental protection cooperation is a major component. Right now, the MEP is earnestly implementing President Xi's proposal announced in the Belt and Road Forum for International Cooperation to set in place an Environmental Big Data Service Platform and an International Coalition for Green Development on the Belt and Road.

For China to actively participate in international environmental protection and global environmental governance is not only out of its own needs to implement green development strategy and protect ecological environment, but also a way to take on its due role in the international community in bringing forward a shared future for the mankind. Going forward, China will continue to expand international environmental cooperation and actively engage in global environmental governance by further strengthening South–South environmental cooperation and promoting the BRI. All these efforts, in promoting domestic ecological progress and improving environmental quality, will as well contribute to the UN 2030 SDGs at the regional and global level.

Liu Youbin: Now the floor is open.

People's Daily: The *Minamata Convention on Mercury* is expected to take effect in August. What has the Ministry done to prepare for this? What needs to be further improved? Has the developed countries pledged any support to its implementation?

Guo Jing: Thank you. It's one of the mandates of MEP to implement international environmental agreements. Over the past years, both the governments and the research institutes have worked hard to implement international conventions, making important headways and winning the international recognition. We are very glad today to have my colleague Mr. Yu Lifeng with us to further brief you on the details. Mr. Yu is the Deputy Director General of Foreign Economic Cooperation Office of MEP, also known as the Technology Center for Environmental Convention Implementation.

Yu Lifeng: As you mentioned, the *Minamata Convention on Mercury* will take effect on Aug. 16. China is among the first countries to sign it. I once came across a media report, saying that starting from 2021 the manufacture, import, and export of fluorescent lamp and mercury-containing batteries will no longer be allowed. It's true, but the implementation work is more than that.

Mr. Yu Lifeng, Deputy Director General of Foreign Economic Cooperation Office of MEP

As required by the *Minamata Convention on Mercury*, we've figured out an implementation timetable based on an initial assessment. By 2020, we will stop manufacturing, importing and exporting all mercury-added products including mercury-added batteries and fluorescent lamp that are not registered for exemptions from the phase-out, ban import and export of mercury-added thermometers and sphygmomanometers, and cut the use of mercury in terms of per unit production of VCM products by 50% by the year 2020 against 2010 use. By 2032, we will shut down primary mercury mining. By 2020, we will work out a preliminary inventory of emissions from sources like coal-fired power plants to make clear the key sources of emission to be controlled, and then further set reduction targets and measures.

The MEP has done a lot of preparation work, including efforts to revise and issue together with relevant departments standards for mercury-related emissions, evaluate 10 mercury-related sectors, develop a national strategy and action plan, and carry out bilateral and multilateral cooperation on capacity building of convention implementation.

In the future, we will continue to develop the national strategy and action plan, and take such steps as controlling the source, cutting the total mercury use and emission, and developing alternative technologies to realize a mercury-free for those related industries. Targeted approach will be adopted to reduce mercury emission and release based on different sectors stage by stage. And plans will be formulated to address mercury-containing waste and the contaminated sites step by step.

Given the fact that China is a major country manufacturing and using mercury, it may be objective to say that China is actually under huge pressure in implementing the *Minamata Convention on Mercury*. Yet we are still quite confident in fulfilling the set targets as years of relevant experience in implementing other conventions or agreements has laid a very solid foundation.

First, as early as the 12th Five-Year Plan period, China has already included mercury as one of the five priority heavy metals to be controlled. Some sectors have begun to control mercury pollution even before China joined the convention. It's reasonable to say that the *Minamata Convention on Mercury* enjoys more favorable conditions in terms of implementation in China than other chemicals-related conventions or agreements.

Moreover, in line with the principle of common but differentiated responsibilities, China as a developing country developed its own distinct model of convention implementation based on years of experience in implementing the *Montreal Protocol on Substances that Deplete the Ozone Layer*, the *Convention on Biological Diversity*, and the *Stockholm Convention on Persistent Organic Pollutants*, etc. This would be of valuable reference for the *Minamata Convention on Mercury*.

Here are the major highlights of this model. First, China has a national implementation plan for each convention as an overarching guidance to such efforts. Second, each convention or protocol is supported by a national coordination mechanism that involves concerned departments and local authorities to ensure this covers all sectors and all regions across China. Third, relevant standards and regulations for implementing international conventions or protocols are integrated into the national environmental management system, paving way for the establishment of sound institutions for environmental management.

Implementation of international environmental convention or protocols is a win-win process for both environment and development in China. It is a strong impetus for Chinese industries to phase out backward industrial capacities and improve relevant technologies.

I would like to share with you some important figures.

In implementing the *Montreal Protocol on Substances that Deplete the Ozone Layer*, China phased out over 250,000 tons of substances that deplete ozone layer, taking up over half of the combined total of developing countries.

In implementing the *Stockholm Convention on Persistent Organic Pollutants*, China banned the manufacture, use, import and export of 17 permanent organic pollutions (POPs) including DDTs, brought down emission intensity of dioxin in key sectors by over 15%, and properly disposed 50,000 tons of POPs-containing waste in more than 100 sites. This actually addressed a number of POPs-related pollution that threatens public health.

On the front of *Convention on Biological Diversity*, China has expanded terrestrial protected areas to 1.7 million km^2, or 18% of the national total, realizing the goal of 17% by 2020 specified in this convention ahead of the schedule. Over 90% types of the terrestrial natural ecosystems, 89% types of the wildlife communities under key national protection and most of the key natural relics are well protected in nature reserves.

Speaking of the future targets in implementing the above conventions and protocols, we plan to phase out 93,000 and 80,000 tons of ozone-depleting substances, respectively, in the production and consumption sector by 2020, and the newly added POPs to be controlled including perfluorooctyl sulfonic acid (PFOS), endosulfan, and hexabromocyclododecane (HBCD) by the date specified in the Protocol. We will implement some key projects on biodiversity conservation, survey and assess biological diversity across the country with each county as a unit, establish a system on monitoring, assessment and early warning, and set in place a biodiversity conservation network.

Above is an overview of implementation of international environmental conventions and protocols in China. Thank you.

Financial Times: The *Law of the People's Republic of China on Administration of Activities of Overseas Non-governmental Organizations (NGOs) in the Mainland of China* (hereinafter referred to as the NGO Law) took effect this year. What is your prospect for the development of NGOs, particularly environmental ones in China? How will the review and approval of environmental NGOs unfold next step?

Guo Jing: Since the reform and opening up, especially from the 1980s, overseas environmental NGOs have carried out many activities in China, which in general positively influenced China's ecological and environmental protection. The Chinese government, including the MEP has always been open and welcomes overseas environmental NGOs to operate in China. As is known to all, as approved by the National People's Congress (NPC), the NGO Law entered into force this year, and we will act in strict accordance with this law.

In responding to national call, the MEP set in place a working mechanism where a team was organized to provide professional service for overseas NGOs through a service platform. The general requirements are to examine and approve the operation of overseas NGOs in China in accordance with the law, to better manage their activities and bring their role into full play. To date, we have achieved some milestones. Of those NGOs taking the MEP as their sponsor in the first round of application, two have finished registration with Beijing Municipal Public Security Bureau by the end of this June.

I'd like to take this opportunity to clarify the registration procedures when overseas environmental NGOs apply to take MEP as their official sponsor. First, they have to comply with relevant laws like the *NGO law*, and also meet requirements of the *Guide for the Registration of Representative Offices and Submitting Documents for the Record Temporary Activities of Overseas Non-governmental Organizations* issued by the Ministry of Public Security (MPS), the department mainly responsible for regulating overseas NGOs' activities in China.

Second, those NGOs seeking the MEP as their official sponsor shall have their business activities in line with that of the MEP as specified in the issued *List of Fields of Activity, Categories of Projects, and Official Sponsors for Overseas Non-governmental Organizations (2017)*. The list makes clear which kind of NGO is eligible to seek the MEP as their official sponsor. The mandates of the MEP mainly include environmental pollution prevention and control, and biodiversity conservation, etc.

Considering that environment is a broad concept, we will examine how much the work of NGOs in and outside China is relevant to that of the MEP when reviewing their applications.

Next we will continue to proceed with it in accordance with the law and relevant procedures. We welcome NGOs friendly with China to carry out work here, and enjoy better development within the legal framework. Thank you.

Rossiya Segodnya: Will China take any steps to deal with imported waste and solid waste?

Guo Jing: Imported solid waste that is eligible for raw materials did play a role in mitigating the resource gap in China at a particular stage of development in the past. Yet with the economic and social development, problems gradually surfaced in China as a result of imported solid waste, including environmental pollution and harm to public health, thus strongly opposed by the society.

Some countries transported their own waste to others through various means, even resorting to smuggling. To bring such transported waste under effective control, the international society adopted the *Basel Convention on the Control of Transboundary Movements of Hazardous Wastes and Their Disposal* in 1989, to which China is a contracting party. As stipulated in the convention, the state of exporting hazardous waste must get the consent of the state of import, which is also an international common practice in controlling imported waste. However, it should be noted that a few businessmen in and outside China are still profiting from such illegal import and export of waste, causing a lot of environmental problems and hence must be cracked down on.

The CPC Central Committee and the State Council attached high importance to the management of imported solid waste as it concerns ecological and environmental protection. In this April, President Xi particularly called for a meeting to deliberate and approve the *Implementation of the Reform Plan to Ban Waste Imports*. The meeting stressed the needs to refine the management system of imported solid waste for the security of national ecological environment and the health of our people. It spells out the needs to develop a timetable on the ban of imported solid waste based on different sectors and categories, and revise the directory of imported solid waste under control category by category and in batches. It's encouraged to employ multiple means including legislation, market and administrative regulation to notably bring down both the number and the categories of imported solid waste. It also advocates circular economy to better manage the recycle and reuse of solid waste. That's why we took immediate action to ban the import of 24 categories of solid waste that are thought hazardous to the environment and of great public concern, including waste plastics, unsorted waste paper, waste textile raw materials and vanadium slag, etc. We also reported this list to relevant committees as requested by the WTO on transparency.

In the future, we will fully implement the decisions made by the central government as well as respond to the new demands and new expectations of our people, to take concrete steps to uphold national ecological and environmental security and protect public health.

Liu Youbin: One more sentence. Starting from July 1, the MEP launched a special campaign against industries involving illegal processing and use of imported solid waste. For more information, please refer to the official Micro-Blog and WeChat account of MEP for daily updates.

The Beijing News: President Xi Jinping advocated that "we should pursue a new vision of green development and a way of life and work that is green, low-carbon, circular and sustainable," and called for efforts to "strengthen cooperation in ecological and environmental protection and build a sound ecosystem so as to realize the goals set by the 2030 Agenda for Sustainable Development" at the opening of Belt and Road Forum for International Cooperation held in May. What has been done by the MEP to promote it, and how could investment under the BRI contribute to local sustainable development?

Guo Jing: The BRI echoes perfectly with the 2030 SDGs. Countries along the Belt and Road are mostly the developing ones and emerging economies, thus facing greater difficulties in balancing economic development and environmental protection. Jointly promoting green BRI does not only serve their own development needs, but also an important step towards realizing the UN 2030 SDGs. It's in the interest of all.

China gave high priority to ecological and environmental protection under BRI, with President Xi stressing on many occasions to step up environmental cooperation and jointly bring forward a green BRI. He also announced to establish an Environmental Big Data Service Platform and an International Coalition for Green Development on the Belt and Road in the Belt and Road Forum for International Cooperation held in this May. The MEP then has made the following efforts to promote green Belt and Road and step up environmental cooperation under this framework.

First, the MEP, the Ministry of Foreign Affairs (MFA), the National Development and Reform Commission (NDRC), and the Ministry of Commerce (MOFCOM) have jointly issued the *Guidance on Promoting Green Belt and Road*, laying out the overarching goals and major tasks in promoting green Belt and Road. It spells out measures to strengthen communication and exchanges, provide environmental protection guarantees for overseas investment, establish a platform for protecting green cooperation, and refine relevant policies and measures.

Second, the MEP issued *The Belt and Road Ecological and Environmental Cooperation Plan*, the top design guiding ecological and environmental cooperation under BRI. It specified 58 tasks in terms of connectivity in policy, infrastructure, trade, finance, and people, in an effort to integrate the ideas and practices of China in promoting ecological civilization and green development into various aspects of the BRI.

Third, the MEP is preparing the Environmental Big Data Service Platform, and the International Coalition for Green Development on the Belt and Road in collaboration with the UN Environment. Other platforms are also underway including a Belt and Road Environmental Technology Exchange and Transfer Center to promote pragmatic cooperation, and a Green Silkroad Envoy Programme that has provided training to over 1100 government officials of environmental departments, young people, scholars, and enterprise representatives from developing countries in the past 5 years.

In terms of the Environmental Big Data Service Platform, it's positioned as one that supports BRI countries in green transition, and facilitating green trade, investment, and infrastructure construction. It will be a flagship window on information exchange, and a platform for sharing expertise and technologies and supporting information collection and decision-making, thus ultimately delivering benefits to all BRI countries.

In terms of the International Coalition for Green Development on the Belt and Road, the MEP and the UN Environment, acting upon the call of our leaders, have begun this work based on the *Memorandum of Understanding between the United Nations Environment Programme and Ministry of Environmental Protection of the People's Republic of China on Building a Green "Belt and Road"* signed in 2016. The goal is, first, to build an international platform to share the ideas, policies, and practices of green development of BRI countries to accelerate such progress across the whole region; second, to promote regional and international study and dialogue to provide policy recommendations for green Belt and Road; third, to bring into full play the role of business in industrial capacities and infrastructure cooperation to ensure this comes in a more environmentally friendly manner; and fourth, to raise the awareness of ecological and environmental protection of both the governments and the people of BRI countries, and enhance their capacity in delivering on the UN 2030 SDGs. We want this Coalition highly inclusive and open to all, thus warmly welcome governments, business, and civil society of BRI countries including relevant departments, localities, research institutes, and civil society in China to be part of it.

In terms of guarding against environmental risks of China's overseas investment, the MEP and the MOFCOM have jointly issued the *Guidelines on Environmental Protection in Foreign Investment and Cooperation* in 2013 immediately after the launch of BRI. It will guide the Chinese enterprises in better complying with local environmental standards, regulations and relevant requirements when making an investment there. There are specific stipulations in terms of implementing these Guidelines and promoting corporate social responsibilities in the *Guidance on Promoting Green Belt and Road* I just mentioned.

By the end of 2016, an Initiative on Corporate Environmental Responsibility Fulfillment for Building the Green Belt and Road was announced by the business sector under the support of the MEP, NDRC, and MOFCOM.

Recently we've seen the Chinese enterprises are getting more and more aware of fulfilling their social responsibilities in foreign investment, with many investment projects highly acclaimed by the local governments and people. For example, a Chinese enterprise that contracted for the Cuddalore thermal power project in India has won the environmental protection award from an Indian foundation on promoting full-scale power generation. In the China-contracted Sahiwal coal-fired power project in Pakistan, the total SO_2 and NOx discharge were, respectively, 180 mg/m^3 and 300 mg/m^3, far below the local upper limit. China also undertook the Municipal Sewage Treatment Infrastructure Construction under the Project of China–Belarus Industrial Park for Stage I. It was completed by this May, generally considered the most advanced sewage treatment station with the largest capacity in

Belarus. All these amply demonstrate that most Chinese enterprises have done a good job in complying with local environmental regulations in their overseas investment, and that's why they got widespread recognition from both the local government and the people.

China Business News: China has been advocating and actively participating in South–South Cooperation. What has been done in terms of environmental protection under this framework? And what's the progress so far?

Guo Jing: You did raise a very important question. Much has been done over the past years. We established a China-ASEAN Environmental Cooperation Center under the MEP 7 years ago, especially responsible for promoting South–South Cooperation. It has another name of China Center for Shanghai Cooperation Organization (SCO) Environmental Cooperation. South–South Environmental Cooperation is mainly led by this center, and supported and joined by other departments of the MEP. I'd like to give the floor to Ms. Zhang Jieqing from this center to respond to your question.

Zhang Jieqing: Thanks. For China to engage in South–South Environmental Cooperation is in the interest of everyone, because both China and other developing countries face similar environmental challenges. South–South Environmental Cooperation facilitates experience sharing on environmental protection and sustainable development among developing countries, by which countries, on the one hand, become more capable of protecting the environment and sustaining their development, and on the other hand, get more engaged in international environmental governance. That's why the Chinese government has been placing South–South Environmental Cooperation high on its agenda, and working earnestly on it.

Ms. Zhang Jieqing, Deputy Director General of China-ASEAN Environmental Cooperation Center of MEP

For one thing, we set up a professional institute under the MEP charged with issues related to South–South Cooperation, which is, as Mr. Guo Jing mentioned, the China-ASEAN Environmental Cooperation Center, also known as China Center for Shanghai Cooperation Organization Environmental Protection and Lancang-Mekong Environmental Cooperation Center.

For another, we jointly carried out many activities. Firstly, we have policy exchange with a number of developing countries through such mechanisms as China- ASEAN Forum on Environmental Cooperation, China-Africa Environment Ministers Meeting, and Forum on China-Arab States Environmental Cooperation. This helps countries to understand each other better and to draw upon good experience. Secondly, we develop strategies to promote cooperation in areas of common interest. Together with the ASEAN countries, we introduced a *China-ASEAN Strategy on Environmental Protection Cooperation* and *China-ASEAN Environmental Cooperation Action Plan*, and then developed detailed implementation activities within the framework every year. In terms of protecting the Mekong River, we brought up a *Green Lancang-Mekong Plan*, will implement it step by step. Thirdly, we forged collaboration on environmental protection industries and technologies. Based in existing eco-industrial parks, we set in place several Demonstration Bases for cooperation on environment-related technologies and industries, through which we are able to share and promote China's integrated approach to environmental problems and enhance environmental protection capacity of both China and other developing countries. Fourthly, we conducted joint research. We produced the *China-ASEAN Environment Outlook* together with the ASEAN countries, which discussed environmental issues facing both, and areas of cooperation. We also introduced capacity building programme called Green Silkroad Envoy Programme. In the past 5 years, this programme has held a total of 52 training sessions involving over 1,000 government environmental officials, scholars and young people from more than 80 countries, for the purpose of building awareness and strengthening capacity in environmental management.

In addition, with the 6 million US dollars Trust Fund pledged by the Chinese government to the UN Environment, we developed a wealth of events and activities in areas of green economy and implementation of international environmental conventions to enhance the environmental governance capacity of more than 80 developing countries.

Going forward, we will further step up South–South Environmental Cooperation. As President Xi said in the Belt and Road Forum for International Cooperation, China will pledge a 60 billion RMB of fund to developing countries and international organizations that are involved in BRI within the next three years. In addition, the Chinese government will continue to support the Trust Fund of UN Environment to build up capacities in addressing environmental problems. All these will bring new opportunities for South–South Cooperation. Thank you.

The Paper: My question is about air pollution prevention and control. Has any international joint research conducted by the MEP on causes for air pollutants, and are there any findings so far?

Guo Jing: One of the focuses of international environmental protection is on addressing immediate environmental issues of China. Air pollution prevention and control has always been a priority of the MEP, especially in its cooperation with the developed countries. For example, China and the US began cooperating on environmental issues as early as the 1980s under the *China-US Science and Technology Cooperation Agreement*, which covered areas of air environment monitoring, standards and air pollution control, among others. We also saw a number of cooperation projects between China and Japan, Germany and other countries involving air pollution prevention and control. The following illustrated the contribution of international environmental cooperation in addressing air pollution in China.

The first is on legislation and institutions. China has drawn upon advanced experience from other countries especially the developed ones, which accelerated its own legislation process.

The second is about some basic scientific research and study on the rules or mechanisms. To better address air pollution, the first thing is to identify the causes for it, so that we are able to find the right approach to it. Over the past years, we cooperated with the developed countries, developing many good projects with the support of universities, research institutes, and government authorities. It enhanced our research capacity, for example on the source analysis, and is quite helpful for us to better understand the causes of air pollution as well as the rules and mechanisms behind it.

The third is on the use of advanced technologies. Staring from the 1980s and 1990s, international environmental cooperation began to focus on introducing advanced technologies and equipment for controlling air pollution in China, and tailoring which to China's national conditions. This remarkably enhanced China's capability in developing relevant technologies and equipment such as de-dusting and air monitoring, taking which to a whole new level.

Last, we believe cooperation must be a two way and mutually beneficial one. International environmental cooperation advances China's environmental protection technologies and boosts its environmental industry market. China boasts tremendous market potential for environmental protection. It's estimated by some experts that the annual growth rate of such market was around 15–20% during the 12th Five-Year Plan period. Looking into the 13th Five-year Plan period and beyond, the market for environmental technologies and industries is expected to expand at a rate no lower than that of the 12th Five-Year Plan period. That represents enormous opportunities for foreign investment into China's environmental protection market. China is open to that, and welcomes international business that are technologically advanced and ready to compete with their Chinese counterparts to invest in China. In benefiting themselves, they will as well expedite the pollution prevention and control especially on air in China. Thank you.

National Business Daily: The handout says China has been stepping up international cooperation. Are there any international cooperation on environmental protection technologies targeting those environmental headaches in recent years, and how much does it help with these issues? What is the major weak link in developing environmental technologies, and what are the next-steps for international cooperation in this regard? Do we have any international experience to draw upon in terms of water and soil pollution prevention and control?

Guo Jing: As we all know, the central government has rolled out the Action Plans, respectively, on Air, Water and Soil Pollution Prevention and Control consecutively since 2013, known as "Ten Measures". The MEP is now proceeding with their implementation with full vigor, including efforts to promote relevant legislation and put forth supporting measures. Updates on our work are available on multiple platforms.

Honestly speaking, the pressure laid on us is rather huge, yet we are still confident, and also determined and persistent in achieving these goals. We are confident because the senior leadership in China gave high priority to environmental protection. We must be determined because our leaders require us to implement this in full and there must be concrete results coming from it, so that our people including everyone present here would enjoy a greater sense of fulfillment. We must be persistent because it's a science-based process, thus taking time to tackle air, water, and soil pollution. Take the *Action Plan on Air Pollution Prevention and Control* as an example. We have seen notable progress in implementing this Action Plan over the past 3 years between 2013 and 2016, when international cooperation on environmental protection played a key role in channeling and promoting national inputs into relevant technologies and research to meet the needs for advanced technologies on air, water, and soil pollution prevention and control. National inputs in environment-related research and development are actually growing year on year, and many departments have placed it high on their agenda.

The focus of international cooperation is gradually shifting towards technological cooperation. In the social domain and from a market perspective, there is increasing business cooperation, which articulates the needs for more appropriate guidance from the government. The MEP had detailed regulations on it and also rolled out a raft of policies. I'd like to call on friends from the media to help promote these concepts and documents. As we work together, we will surely move forward the implementation of the three Action Plans on all fronts including science, technology, and the market. Meanwhile, we must never forget the essential role of innovation, as science and technology are usually the most fundamental approach to many problems, especially to environmental issues.

In terms of weak links, first I have to say China has made notable progress over the past years in advancing technologies for controlling conventional pollution. In a comparative analysis, China has made great strides compared with itself in the past. Compared with BRI countries and the developing ones, China is catching up and even doing better, with many advanced technologies, and equipment for controlling conventional pollution exported to other countries. Yet despite the progress, we still lag behind and lack experience in certain areas. In terms of soil pollution prevention and control, remediation of contaminated sites is one of the big headaches not only facing China, but also the whole world. The US started to tackle soil pollution in the 1960s and 1970s, when the Super Fund Programme was introduced and approved by the US Congress. It requires the polluters to pay for the cleaning of contaminated sites. This is a good piece of experience for us to draw upon, yet we still face difficulties given the different national conditions. I believe one day, with inputs from our own team, China will definitely have its own tailor-made technologies to address air pollution.

China Daily: My question is about international cooperation on air pollution prevention and control. The Tripartite Environment Ministers Meeting among China, Japan and ROK (TEMM) is held every year. Could you brief us on the cooperation among the three countries concerning environmental pollution prevention and control particularly on air?

Guo Jing: China, Japan, and the ROK are important regional partners of each other, and maintained sound environmental cooperation. 18 years ago, the three countries jointly established the TEMM mechanism. To date, 18 meetings have been held consecutively, and the next one will soon be convened. Under the TEMM, the three countries put forward and implemented the *Tripartite Action Plan on Environmental Cooperation* in areas of biodiversity conservation, air and water pollution prevention and control, ecological restoration, public awareness, chemical management, and transboundary movement of electronic waste, etc. Thanks to years of efforts, the three countries basically set in place a relatively sound communication and cooperation mechanism, which are rarely seen elsewhere. Next year, the TEMM 20 will take place in China, when we will draw up a blueprint for the next decade built on the progress achieved in the past 20 years of collaboration among the three countries.

Environmental cooperation for the three countries has come increasingly down-to-earth. A good example is the air pollution you just mentioned, which concerns all of the three countries. Under the TEMM, a Tripartite Policy Dialogue on Air Pollution was introduced and usually takes place in China, Japan, and ROK in turn. Up to now, four dialogues have been held. There are two working groups set up under this dialogue, including one on Scientific Research on Prevention and Control, and another on Technology/Policy on Air Quality Monitoring and Prediction.

Environmental pollution, in a final analysis, is an issue brought about by a country's own development. Both ROK and Japan, and even other Western countries are struggling with air pollution and issues generated from such development. In my opinion, the key is to facing up to these challenges together by strengthening communication and exchanges, enhancing mutual understanding and stepping up cooperation. Under the bilateral cooperation mechanisms, China is able to maintain close communication with Japan and ROK, and learnt a lot in preventing and controlling air pollution in China. As I always mentioned with my Japanese and ROK colleagues, China's success in combating environmental pollution is in itself a great contribution to regional and global environmental protection and sustainable development. If China tackled its own environmental issues, that would address concerns of its 1.4 billion people, namely 1/5 of the world population, which is no doubt a great contribution to the whole world.

China has declared wars against pollution, fighting with full determination to defend the blue sky. If China delivered on the goals set for the coming 5–10 years in improving air quality in time, it will contribute substantially to the regional air quality. I believe my Japanese and ROK colleagues fully understand and appreciate this. Thank you.

Nanfang Metropolis Daily: There's an Environmental Big Data Service Platform to facilitate BRI development. At the same time, there are also some international concerns over the Chinese enterprises that may fail the environment impact assessment or follow less stringent environmental standards. Are there any guidance from the MEP to help these enterprises in environmental compliance?

Guo Jing: I already mentioned the MEP and MOFCOM jointly issued the guidelines on it three or four years ago, which is the policy guidance to these enterprises. Chinese enterprises that follow through the national guidance better will be able to pursue their overseas investment and other business better. It's by its natural in line with the internationally accepted territorial principle for enterprises to abide by the local environmental laws when making investment and doing business overseas. Most enterprises are increasingly aware of the importance of fulfilling corporate social responsibility and respecting local environmental laws when operating overseas. Just now I have shared with you a lot of good examples, and many of our enterprises are doing quite well. It's actually the shared nature of human beings to keep a low profile when doing good deeds instead of broadcasting it. There's also an old Chinese saying that bad news travels fast. We did see some enterprises failed to fully understand the local environmental laws and requirements and thus led to certain problems. I hope they will on the one hand learn the lessons and set good examples for others, and on the other hand share our ideas and best practice with other countries, especially those face similar challenges with China and are less developed, so that they could better protect their local environment. Thank you.

Liu Youbin: That's the end of today's press conference. Thank you.

Regular Press Conference (August)

(August 22, 2017)

The regular press conference of MEP in August was hosted on the morning of August 22, where Mr. Bie Tao, Director General of the Department of Policy and Regulation of MEP, briefed on environmental laws and regulations as well as relevant economic policies, and Mr. Liu Youbin, Counsel (Director-General level) of the Department of Communications and Education, reported on recent progress in prioritized areas in environmental protection and responded correspondents on questions of interest. Implementing *Plan of Optimizing Regular Press Conferences*, this marks the first time that a monthly report of prioritized tasks of the MEP was delivered during the press conference, where environmental issues of public concern were responded.

Ministry of Environmental Protection of the People's Republic of China,
2017 Press Conference Records of Ministry of Environmental Protection, the People's Republic of China, https://doi.org/10.1007/978-981-13-7330-5_9

MEP August Press Conference (1)

MEP August Press Conference (2)

Liu Youbin (Moderator): Dear friends from the media, good morning! Welcome to the August session of regular press conference of the Ministry of Environmental Protection, where Mr. Bie Tao, Director General of the Department of Policy and Regulation, Ministry of Environmental Protection will present on the policy and regulatory development for environmental protection in China, after which he will take the questions from the media.

Besides, starting from today's press conference, we shall update you on the general progress in environmental protection while reporting on the prioritized areas of your keen interest on a monthly basis so as to give you a more holistic understanding of our missions and tasks. To begin with, I would like to give you a general briefing.

1. Central inspections of environmental protection have covered all 31 provinces (municipalities and autonomous regions).

With the 4th round of central inspection group of environmental protection starting their work recently, central inspections of environmental protection have covered all 31 provinces (including municipalities and autonomous regions) in China.

Central inspections of environmental protection focused on salient environmental issues arousing concerns of the central government and public complaints while incurring appalling social consequences. During the pilot inspection in Hebei Province and the first three rounds of inspections, the central inspection group of environmental protection heard over 90,000 complaints from the public, of which 63,260 cases were handed to local authorities after removing repetition and 298 cases of ecological and environmental damage in which governmental or Party officials were held accountable were handed to a total of 23 provinces (municipalities and autonomous regions). Up to now, the complaints having been heard and investigated are basically concluded with nearly 12,000 individuals held accountable, solving many environmental issues arousing public concern, which was applauded and supported by the public. The procedures for accountability investigation have been completed for the 100 cases of ecological and environmental damage handed over during the first round of inspection, and the results will be publicized soon.

Currently, the 4th round of central inspection of environmental protection is underway. Up to August 20, 2017, a total of 10,268 complaints were heard with 938 individuals held accountable.

2. The Beijing–Tianjin–Hebei Region and Its Periphery Areas shall take action for comprehensive air pollution control

Since April 7th, the intensified inspection on air pollution control in Beijing–Tianjin–Hebei region and its periphery areas has been conducted for over 4 months with 9 rounds of inspections concluded. Feeling the pressure from the unyielding and rigorous law enforcement, enterprises have seen substantial improvement in compliance rate in emissions as initial achievements are made in cleaning up enterprises that are in unauthorized locations, lack proper certificates and fail to meet emission standards, and the aggravation of environmental pollution is effectively curbed. Up to August 20th, intensified on-site inspections were carried out in 40,925 enterprises, of which environmental problems were found in 22,620 enterprises; 9,402 outstanding issues were handled under supervision and MEP has had admonitory talks with the responsible officials of 8 cities (counties, districts).

Going forward, more targeted measures are to be taken to tackle the weak links in air pollution control in autumn and winter, with "Action for Comprehensive Air Pollution Control in Beijing–Tianjin–Hebei region and its periphery areas" in place. The "1+6" Work Plan has been approved by the Party Member's Group of the MEP upon discussion, namely 1 master plan plus 6 supporting plans, featuring holistic planning throughout the institutional system, focusing on key areas and key industries, coordinating inspection and follow-up inspection, urging the local authorities to disclose information in a timely manner and implementing quantifiable accountability in order to consolidate the results of inspection, prevent the relocation or reopening of enterprises that are in unauthorized locations, lack proper certificates and fail to meet emission standards and make down-to-earth efforts in air pollution control in autumn and winter.

3. Further clear up and reorganize imported solid waste treatment and reuse industry

To thoroughly execute the decision and roadmap of the CPC Central Committee and the State Council for strengthening eco-environmental protection, the MEP has launched a targeted action to crack down illegal conducts in imported solid waste treatment and reuse since July 1st for one month.

The crackdown action mobilized 1,700 inspectors for law enforcement inspection on 1,792 enterprises engaged in imported waste treatment and reuse throughout the country, screening and cracking down violations of environmental laws and regulations. Incompliance or violations were found in 1,074 enterprises, or 60% of all enterprises inspected during the inspection.

To further implement the State Council's requirements on banning imported trash and strengthening reforms of solid waste management system, the MEP has identified the following four areas for the next step: first, deal with violations of environmental laws and regulations spotted during the inspections with the results disclosed in public. Second, make amendments to Catalogue of Imported Waste Management. The MEP, in collaboration with other relevant authorities, has recently published Catalogue of Imported Waste Management (2017), re-categorizing 24 types of solid waste of 4 categories from Limited Import to Prohibited from Import, which is to come into effect by the end of 2017. Third, launch special clear up action of imported solid waste hub. A 5-month rectification and clear up of recycling industry shall be jointly initiated with other relevant authorities with hubs of treatment and reuse as the highlighted area, so as to eradicate illegally imported trash and provide guidance for the healthy development of waste recycling businesses. Fourth, start the revision of environmental protection standards for imported solid waste usable as raw materials. Joint research and discussions are currently underway and the revision shall be announced in public very soon.

This afternoon, Mr. Li Ganjie, Minister of Environmental Protection shall host a workshop with relevant governmental officials of provinces and municipalities as well as responsible persons of the State Council to jointly discuss on strengthening the ban on imported trash and improving solid waste management.

4. Strengthen efforts in implementing the *Action Plan on Water Pollution Prevention and Control (Action Plan on Water)*

Generally speaking, positive progress has been made nationwide in water pollution control; yet certain regions and industries are still lagging behind. From January to July 2017, 70.2% of surface water has reached Water Quality Grade III or above, a year-on-year increase of 1.1 percentage points, while 8.6% of water bodies are below Grade V, a year-on-year decrease of 1.2 percentage points. Up to July this year, water quality remediation program has been completed in 364 out of 681 identified black and putrid water bodies in key cities, accounting for 53.5% of the total. In terms of industrial pollution control, 1,771 enterprises have completed clean production renovation, accounting for 85.0% of the total; 417 industrial hubs of or above provincial level, or 18% of the total, have not built wastewater collection and treatment facilities as required. In terms of urban water pollution control, 173 urban water treatment facilities in sensitive areas have reached Level I discharge standard for water pollutants, accounting for 61% of the total; and regarding rural water pollution control, 14 provinces (autonomous regions) have not yet completed restrictive zoning for animal husbandry.

For the next step, we will focus on the following tasks.

First, improve the comprehensive supervision mechanism to provide guidance for environmental authorities at all levels in building a water environment analysis system, identifying underperforming regions and most evident water environmental issues, and implementing comprehensive supervision measures for most challenged regions in a coordinated manner. Second, strengthen the supporting policy measures with accelerated efforts in regulations, standards and guidelines of drinking water source protection, underground water pollution control, etc. Third, organize midterm evaluation by an expert panel to assess the water environment quality improvement since the implementation of the *Action Plan on Water*, identifying the proposed approaches and prioritized tasks in the upcoming period. Fourth, redouble efforts in information disclosure. The MEP has recently reported the progress in the targeted tasks described in the Action Plan as well as relapse in water quality in all provinces in the first half of 2017. The MEP shall continue our efforts in information disclosure and subject ourselves to public supervision.

Now I would like to invite Director General Bie Tao to update you on the progress in our work.

Bie Tao: Good morning, everyone! First, thank you for your interest in and support for environmental protection, including the policies, laws, and regulations. As arranged by the Ministry, I would like to brief you on the progress in environmental laws and regulations as well as relevant economic policies since the 18th CPC National Congress.

In recent years, new trends are observed in environmental laws and regulations as well as environmental economic policies. President Xi Jinping emphasized during the 6th collective study session of the Political Bureau of the CPC Central Committee on May 24th, 2013 that "only by implementing the strictest regulations and the most rigorous rule of law can we ensure sound ecological development," which creates a rare historical opportunity for strengthening environmental policy-making and regulatory work.

Since the 18th CPC National Congress, positive progress has been made in environmental policy-making and regulatory work on national level, as is demonstrated in the following four dimensions.

Bie Tao, Director-General of the Department of Policy and Regulation, MEP

1. Accelerate legislation in environment with 7 laws promulgated within 5 years.

Since 2012, the MEP has been closely collaborating with legislators in the formulation and revision of 7 environmental protection laws while facilitating the formulation and revision of 9 administrative regulations and 22 departmental regulations concerning environmental protection. Currently, the Ministry is collaborating with the legislative authority of the National People's Congress in developing nuclear safety law and law on the prevention and control of soil contamination while working with the Ministry of Finance in formulating the regulations on implementation of Law on Environmental Protection Tax. Evident progress is seen in environmental legislation while environmental protection is one of the most notable field of legislation in terms of dynamism in legislation activities, frequency of law revision, outstanding accomplishments and well-structured system, and legislation in ecological and environmental protection represents an important component of the socialist legal system with Chinese characteristics.

2. Give equal importance to making new laws and revising existing laws, improve legislation quality and fill the gaps.

First, gaps in legislation are filled. *Law on Environmental Protection Tax* is the first single tax law in accordance with the principle of "collecting tax based on law" while reflecting tax and fees reform and environment-friendly tax system. The formulation of *Nuclear Safety Law* and *Law on Soil Contamination Control* has filled in gaps in legislation.

Second, responsibilities are reinforced for each stakeholder, with clarification in responsibilities, rights, obligation, and assurance mechanism of government and its agencies, enterprises, and the general public in environmental protection.

Third, deterrence of law is strengthened through higher cost of violation. In terms of administrative responsibility, no ceiling is defined for punishment, with additional means of punishment such as consecutive punishment by day, seal-up and detention, transfer for detention and fine of 1%-5% of total investment for projects launched into construction without environmental approval. Regarding criminal liabilities, the MEP has been collaborating with the Supreme People's Court and the Supreme People's Procuratorate in developing and revising Judicial Interpretation of Environmental Pollution Crime, adopting more powerful measures in cracking down such crimes. In terms of civil liabilities, articles concerning liability in environmental tort and public interest litigation initiated by social organizations based on law are added.

Fourth, accelerated progress is seen in the formulation and revision of laws and regulations urgently in need for deepened reforms in all aspects. For example, *Law on Environmental Impact Assessment and Management Regulations for Environmental Protection of Construction Projects* are revised; items for environmental approval are further simplified for construction projects with strengthened in-process and after-event regulation. Approval is replaced by registration for Registration Form for Environmental Impact of Construction Projects (Registration Form for Environmental Impact was required for 50% of construction projects in the past), while administrative approval for completion acceptance for environmental protection facilities is replaced by self-inspection and acceptance by enterprise, of which procedures and criteria are stipulated by the MEP. Industrial prequalification is abolished and serial examination and approval are replaced by parallel examination and approval, while approval of environmental assessment is no longer a prerequisite for approval of investment projects or a procedure for industrial and commercial license. Each governmental agency shall take up its respective responsibility and pursue higher efficiency to reduce burden for enterprises.

3. In legislative system, state laws and Party rules and regulations are mutually complementary and reinforcing. Party rules and regulations constitute a major component of socialist legal system. The MEP is actively involved in and coordinating with the Party in formulating Party rules, regulations and policy documents concerning environmental protection, with much importance attached to the synergy between state-level environmental legislation and Party rules and regulations for environmental protection. At least, 20 Party rules, regulations, and policy documents for environmental protection are promulgated upon discussion and approval by the Central Leading Group for Overall Reform, which mainly include *Masterplan for Institutional Framework for Promoting Ecological Progress, Measures of Holding Party and Governmental Leaders Accountable for Eco-Environmental Damage, Plan for Environmental Protection Inspection, Pilot Plan for Compensation System Reform for Eco-Environmental Damage, Implementation Plan for Controlled Pollutant Emission Permit System*, etc.

4. Create the synergy of legal restraints and policy incentives for better results. While enhancing the environmental legislation system and strengthening the mandatory legal restraints, efforts are redoubled to accelerate the development and implementation of environmental economic policies to encourage eco-friendly production and consumption. First, press ahead with green financial reform to innovate the mechanism for green development. The MEP is involved in publishing *Guiding Opinions for building a Green Financial System* and implementation plan for pilot zone of green financial reform and innovation in five provinces and autonomous regions. We have initiated *Administrative Measures for Compulsory Liability Insurance for Environmental Pollution*, put in place regulatory framework for environmental information disclosure by listed enterprises, made progress in green credit and revised green bond catalogue instructions. Second, facilitate tax system reform to encourage enterprises to expand investment in environmental protection. The Ministry pushes forward the creation of environmental protection tax and promotes green initiative in existing tax categories such as consumption tax, corporate income tax, value added tax, etc., granting tax incentives for enterprises conscientiously investing in environmental protection with effective emission reduction according to law, while imposing higher consumption tax on products incurring heavy pollution so as to encourage green consumption. Third, establish an environmental credit system to guide enterprises to improve their environmental performance. Environmental credit indicator system has been continuously improved with credit information collection and interdepartmental communication mechanism reinforced. Enterprises with good environmental credit are entitled to preferential policies while those with poor environmental credit are punished by 31 departments.

Now I would like to take your questions.

Science and Technology Daily: Recently disputes are aroused by the "one-cure-for-all" policy in pollution control in some places. What is your opinion about this problem?

Bie Tao: This problem has aroused much public attention, and your question reflects how people are concerned about environmental law execution. Up to now, four rounds of central inspections of environmental protection have been carried out, covering all 31 provinces (including municipalities and autonomous regions) in China. The shift from enterprises to government as the target of supervision has brought major impact on environmental regulation. Reinforced inspection on key regions such as the Beijing-Tianjin -Hebei region has produced positive results, yet problems are also spotted. According to statistics of the MEP, violations are found in over 170,000 enterprises in the Beijing–Tianjin–Hebei region, which is indeed alarming. Be it government or enterprise, the supervision is open and transparent and report by public is encouraged so that all irregularities shall be exposed. I think this is a positive phenomenon.

From a legal perspective, problems occurring in enterprises include construction without approval, inconsistence between approved project and constructed project, environmental protection facilities with no prior approval or failing to operate, excessive discharge of pollutants, absence of discharge permit, delay in or refusal to implement rectifying measures after being ordered by environmental agency to make rectification. Various forms of violation are spotted and should by no means be tolerated. Corresponding punitive measures are stipulated in laws and regulations for enterprises breaching the law. Construction without approval, failure of environmental protection facilities and excessive discharge of pollutants are, respectively, subjected to specific punishment. For example, unapproved construction projects may be suspended by environmental agencies and fined; excessive discharge of pollutants may be subjected to production restraint or suspension for rectification, and even closedown upon government approval in severe cases; business owners refusing to obey the order by environmental agency to stop the violation may be handed to public security department for administrative detention. Some acts of breaching, turning a blind eye to laws and regulations, aim at pursuing maximum profit at the cost of public environment and fair play of the market. The cost of environmental protection for some enterprises that are in unauthorized locations, lack proper certificates and fail to meet emission standards is next to zero, which means that the actual cost of the same product is much lower for offenders. This is unfair for their competitors who obey the law. Therefore, rigorous law execution, or zero tolerance, is a must for the offenders, which is essentially required by law-based governance.

Rigorous law execution is an essential principle of rule of law, eliminating the "narrow escape" mentality of enterprises while granting law-abiding players a fair market environment so that no loss shall be suffered by law-abiding enterprises, nor would violators gain any profit from breaching the law. The MEP never required local environmental agencies to seek "one-cure-for-all"; instead, we always remain clear in two things. First, we are opposed to the nonfeasance and negligence of supervision of certain local governments which caused offenders and pollution to exist for a prolonged period of time; second, we are opposed to the practice of nonfeasance as a routine and handling the relationship between development and environmental protection in a simplified and brutal manner at the time of environmental protection inspection and investigation. This is very irresponsible and abusive. We are opposed to nonfeasance as a routine as well as power abuse at the time of inspection.

From a legal perspective, environmental supervision and law execution should follow the following three principles or standpoints. First, zero tolerance for offenders, rigorous and fair execution of law; second, fair treatment and protection of legitimate rights of business operations for enterprises obeying environmental protection laws; third, categorized management and rational guidance for law violators based on conditions and procedures stipulated in law, administration by law on a sound and reasonable basis. This is my viewpoint on the "one-cure-for-all" issue.

Reuters: We would like to understand how to build a long-term and sustainably effective inspection mechanism. Is China aiming at adopting more legal measures than administrative measures? Is the existing legal system sufficient to deal with such an enormous number of cases?

Bie Tao: An increasing number of companies violating the law are exposed with reinforced measures in environmental protection inspection. Before the inspection in April, self-assessment was carried out on a local basis in Beijing–Tianjin–Hebei region, during which 50,000 to 60,000 enterprises were reported to have violated the law. This result was shocking enough. However, by the end of June, according to the statistics from Environmental Inspection Bureau of the MEP, 176,000 problems were exposed during the reinforced inspection in the Beijing–Tianjin–Hebei region, approximately 120,000 more than the number based on self-assessment, which shocked us even more. This shows that our measures of law execution and supervision are effective in terms of innovation, environmental protection inspection and follow-up inspection, etc., as many hidden problems and law-breaching practices are exposed.

The second question is how to maintain the positive momentum of inspection and follow-up inspection with lasting effects. We need to build upon the rule of law, which is essentially based on four prerequisites, i.e., legislation with a scientifically sound basis, rigorous execution of law, judicial fairness and law-abiding citizens according to *Decisions on Several Major Issues in Promoting the Rule of Law* by the CPC Central Committee. Spotting the problems is the first step, which should be followed by solving the problems through rigorous administrative execution according to the laws in force. In this process, some issues would be brought to court for lawsuit and resolved through fair judicial procedures; finally, we shall encourage the public and enterprises to abide by law with self-consciousness. This, I believe, is an ideal state and is our goal. I believe that the solution essentially lies in strong rule of law with sound legislation, rigorous execution, judicial fairness and citizens abiding by law with efficiency, self-consciousness, and civility.

The current environmental legal system is effective but not without deficiency. As the administrative authority, the MEP has built up a high-pressure momentum in cracking down environmental violations with the leadership of the CPC Central Committee and the State Council, the guidance and support of the Standing Committee of National People's Congress, the strong cooperation of the Supreme People's Court, the Supreme People's Procuratorate, the public security and judicial authorities as well as the supervision from media and public. We have introduced many mechanisms including punishment on enterprises, administrative detention and criminal liability of accountable persons, etc., and those who fake monitoring statistics or illegally discharge pollutants may end up in prison; such are by no means rare cases. It can be safely stated that environmental protection authorities are implementing every possible measure and mobilizing all possible resources to crack down law-breaching activities. Inspections on local Party Committee and government in the name of central government are carried out to tackle failure in performing environmental protection responsibilities by relevant authorities, enterprises and local governments. This is innovation in mechanism. We are continuously improving environmental damage assessment mechanism with a set of assessment methods established. Cases where criteria of criminal cases are met based on assessment result are handed over to judicial authorities by environmental protection department to hold law-breakers accountable according to law. Besides, environmental protection departments are actively facilitating information disclosure and expanding media coverage, guiding NGOs to engage in environmental issues to strengthen their supervision of environmental protection efforts, while social organizations specializing in environmental protection are encouraged and supported to file public interest lawsuits according to law.

From a positive perspective, the existing legal system is effective but not without weaknesses. Had the legal system been effective enough, there would not have been such an enormous number of enterprises violating the laws and regulations in the Beijing–Tianjin–Hebei region and its periphery areas. The fact indicates that the institutional measures of the current legal system are partially defective. In the future, we will continue with information disclosure and public participation, motivating the entire society to crack down environment-related illegalities with joint efforts.

Besides, we are pressing ahead with institutional and mechanism reforms in environmental protection. For example, through vertical administrative institutional reform of environmental monitoring, inspection, and law execution, the limited payroll of environmental authorities become front-line law enforcers, and efforts are made in establishing dispatched agencies in key regions and river basins. Improvement in institutional mechanism is a gradual process while cracking down illegalities requires a process. Improvement in environment quality is a challenge faced in a certain phase of economic and social development. Since the London Fog incident, it took UK no less than 30 years to remediate and improve its environment. Therefore, environment quality improvement is a relatively long process and we all want to make it shorter, defending the blue sky and maintaining a pleasant environment with green mountains, clear water, and vegetated land. This is our common aspiration. Thank you.

People's Daily: How is the progress of environmental credit assessment in China now? What measures are being taken in joint punishment? What improvements are expected for the next step?

Bie Tao: Environmental credit assessment is something quite new. Environmental protection authorities conduct an overall assessment of enterprise's environmental compliance based on environmental policies, laws, and regulations and the results are open to public. Enterprises and consumers with high awareness for corporate social responsibility of environmental protection may differentiate their choice based on the results and opt for products by enterprises with good environmental credit, hence motivating enterprises to improve their environmental performance.

In recent years, precious opportunities are seen in environmental credit assessment, as the CPC Central Committee and the State Council has attached great importance to the building of social credit system since the 18th CPC National Congress, with multiple guiding opinions and other documents published and circulated for systematic deployment, showing the direction for the environmental credit system building.

Environmental credit assessment is essentially a means of interim and ex post regulation. According to the trend of environmental regulatory reform, completion of environmental assessment is a fundamental prerequisite for legitimacy of construction projects invested by enterprises, while corporate environmental credit assessment mechanism is introduced in interim and ex post regulation by environmental protection authorities after launching into construction and operation. To this end, the MEP has initiated the publication of *Measures of Corporate Environmental Credit Assessment (Trial)* in 2013 and *Guiding Opinions for Strengthening the Construction of Corporate Environmental Credit System* in 2015 to guide local governments to conduct such assessment.

According to *Measures of Corporate Environmental Credit Assessment (Trial)*, key pollution sources or enterprises discharging toxic or harmful substances or located in environmentally sensitive zones are included in the assessment by environmental protection authorities; based on assessment indicators, enterprises are categorized into four grades and are marked with different colors. Proceeding from local situation, practices with local characteristics were established in Jiangsu, Guangdong, Hunan, Sichuan, etc.

Taking Jiangsu Province as an example, I would like to elaborate on how it works. The assessment results in Jiangsu are classified into five grades: the first grade, that is, the top of the "olive," refers to enterprises with strict compliance with environmental regulations and excellent environmental management, which are rated green enterprises with highest environmental credit; the second grade which includes most of the enterprises is known as blue enterprises, referring to enterprises complying with laws and regulations but not yet achieving green criteria; the third grade is known as yellow enterprises with certain problems in environmental management. The fourth grade, rated red, refers to enterprises failing to comply with regulations with evident environmental illegalities. The fifth grade refers to enterprises with very poor environmental performance, which are rated as black enterprises. Red and black enterprises compose the lower end of the olive-shaped structure.

In Jiangsu Province, green enterprises are entitled to preferential policies in finance and pricing, while red and black enterprises are subject to differentiated power bills (5 to 10 cents higher than market price), sewage treatment fees and other punitive measures. Therefore, a major power-consuming enterprise may need to pay millions or tens of millions higher due to differentiated power price due to poor environmental performance and therefore poor credit rating. This has converted the external environmental cost to internal fees and charges payable by enterprise.

Besides, I believe that the corporate image is demonstrated to the society through assessment results, enabling joint binding force from multiple authorities and motivating consumers to make green choices. Consumers are more willing to purchase products manufactured by environmentally- friendly enterprises even if they have to pay more for environmental protection, and many consumers are likely to avoid products from enterprises with poor environmental performance, hence enterprises with good environmental performance would gain more market space.

We will continue to polish these measures, making the more precise and transparent, and involve more stakeholders in assessment to make results more sound easier to understand by enterprises and general consumers so to help them make better informed choices.

To sum it up, environmental credit assessment is a very important mechanism as it plays a crucial role in complementing the mandatory restraints in environmental protection. Thank you.

Beijing Evening News: Could you please give an overall introduction of environmental legislation since the 18th CPC National Congress? In particular, would it be possible that the reform plan of ecological environment damage compensation system be brought onto the legislation level?

Bie Tao: I would like to start by answering your first question. Thank you for your interest in environmental legislation, and I would take this opportunity to give you a brief introduction of environmental legislation. At the expiration of the term of office of the 11th NPC, the fundamental conclusion was that the socialist legal system with Chinese characteristics has been shaped as planned. And the legal system was further improved through promulgation of new laws and revision, abolishment and interpretation of existing laws during the term of the 12th NPC. In line with the requirement to build a systematic and integral institutional framework for eco-protection stated during the 3rd Plenary Session of the 18th CPC National Congress and the mandate to protect the eco-environment with strict legal system during the 4th Plenary Session of the 18th CPC National Congress, we have collaborated with the legislators to revise the fundamental legislation in environmental protection and strengthened the legislation in key areas such as atmosphere, water and soil. The successive formulation (revision) and promulgation of a number of key laws and regulations have provided a strong guarantee for ecological civilization building and environmental protection undertakings.

On the level of law formulation (revision) by National People's Congress, environmental protection authorities have collaborated with legislators in formulating (revising) seven laws, including *Environmental Protection Law* (2014), *Law on the Prevention and Control of Atmospheric Pollution* (2015), *Law on the Prevention and Control of Water Pollution* (2017), *Law on the Prevention and Control of Environmental Pollution by Solid Waste* (2016), *Marine Environment Protection Law* (2016), *Environmental Impact Assessment Law* (2016), *Environmental Protection Tax Law* (2016). And we are currently collaborating with legislators in the formulation of Nuclear Safety Law and Law on the Prevention and Control of Soil Contamination.

On the level of administrative regulations, we are actively facilitating the formulation and revision of nine administrative regulations, including *Regulations on the Prevention and Control of Pollution from Large-scale Breeding of Livestock and Poultry, Regulations on Urban Sewage Discharge and Treatment, Measures for the Administration of Permit for Operation of Dangerous Waste, Regulations on the Administration of Prevention and Control of Marine Environment Damage by t Engineering Project Construction, Regulations on the Administration of Prevention and Control of Marine Environment Damage by Marine Engineering Project Construction, Regulations on the Administration of Prevention of Marine Environment Pollution from Vessels*, and *Regulations on the Administration of Environmental Protection for Construction Projects* which has been recently revised.

On the level of departmental rules and regulations, in 2015, in support of the execution of Environmental Protection Law and other related laws, important rules formulated by environmental protection authorities include *Measures for the Restriction and Suspension of Production by Environmental Protection Authorities, Measures for Consecutive Punishment on Daily Basis by Environmental Protection Authorities*, etc., and *Measures for Administration of Soil Environment of Contaminated Land Parcels* in order to implement the *Action Plan on Soil* published by the State Council. Recently there are several other rules and regulations being drafted, including measures for the administration of environmental protection of farmland, etc. 22 departmental rules and regulations have been promulgated during the past four or five years.

Besides, the MEP has been actively involved in the formulation of some Party regulations and policy documents, including *Measures for the Accountability of Party and Government Leaders for Damage to the Ecological Environment, Plan for Environmental Protection Inspection, Pilot Plan for Reform of Ecological Environmental Damage Compensation System, Implementation Plan for Permit System of Pollutant Discharge Control*, etc.

About the inclusion of ecological environmental damage compensation into law, the task of constructing such a system was stated in the *Decision of the CCPCC on Some Major Issues Concerning Comprehensively Deepening the Reform. Pilot Plan for Reform of Ecological Environmental Damage Compensation System* was published by the General Offices of CCPCC and the State Council upon approval by the Leading Group for Overall Reform in November 2015. In 2016, upon approval by the Central Government, regional implementation plans were successively published in 7 pilot provinces and municipalities including Jilin Province. Follow-up and periodic review and evaluation were carried out on the performance of the pilot zone by the MEP at the end of this June. Seen from the current situation, the pilot project is basically progressing well and some principles have been shaped by trial and error. As is required by the Central Government, the ecological environmental damage compensation system shall be expanded nationwide for trial implementation from 2018. We have already drafted a masterplan for such nationwide trial implementation based on experience drawn from the pilot zones and the plan has been reported to the State Council.

According to the current laws, victims may resort to civil litigation for physical or property damage by enterprises discharging pollutants. The object for ecological environment damage compensation is public, or state-owned rights, and such right cannot be claimed by average individuals. In the case of damage to state-owned resource elements by pollution, the State Council, representing the collective ownership, may authorize the provincial government, which shall in turn appoint competent authorities as the right holder to claim for compensation. There are two possibilities in claim, one being consultation on equal terms; the other being litigation for ecological environmental damage compensation should the consultation fail. As for granting the status of law, it should be noted that we are currently at the point of pilot reforms, and the MEP has reported to the Supreme People's Court, the Supreme People's Procuratorate and the legislators on problems arising from reality, aiming at promulgating some specified procedures and rules based on the progress of the pilot projects to ensure effective implementation of the reforms. Of course, we do not exclude the possibility of formulating laws and regulations correspondingly, depending on how the pilot project works.

Xinhua News Agency: The MEP signed an agreement with National Securities Regulatory Commission in June for the supervision of environmental information disclosure of listed companies. How is it progressing now? What steps are to be taken next?

Bie Tao: In August 2016, prior to Hangzhou Summit, the Leading Group for Overall Reform has approved *Guiding Opinions on Accelerating the Building of a Green Financial System*, clearly stating that the environmental information disclosure for listed companies shall be further facilitated. In order to implement the Guiding Opinions, the MEP together with the National Securities Regulatory Commission, signed the Cooperation Agreement as you have mentioned. According to this Agreement, the two agencies will first facilitate disclosure of environmental management information in annual report of listed companies and their subsidiaries and some progress have been made. I would like to elaborate case by case.

The first case applies to over 160 listed companies categorized as "enterprises under special state supervision" by the MEP among more than 3,000 listed companies in Shenzhen and Shanghai Stock Exchanges, nearly 90% have disclosed their environmental information, of which we are basically satisfied. Generally speaking, these 160 companies are operating well with large economic output and considerable environmental impact; public supervision through information disclosure as a means to improve their environmental performance plays a positive role in overall environmental improvement. However, some problems do exist as some enterprises fail to disclose environmental information as required, or some information is found to be incomplete or incompliant with requirements, or concealing existing problems while missing the details. We are collaborating with the National Securities Regulatory Commission to take corrective measures continuously.

The second case applies to listed companies categorized as enterprises under special supervision by provincial or municipal environmental protection authorities. These companies are also required to disclose their environmental management information. The list of companies is being screened and verified now.

The third case applies to where at least one major subsidiary is categorized as enterprises under special supervision by environmental protection authorities on various levels, of which environmental information must be disclosed by their listed parent company. Currently, companies listed in Shanghai and Shenzhen Stock Exchanges have approximately 70,000 subsidiaries, and we are comparing and screening them one by one.

According to the securities regulatory requirements, 2017 semiannual report must be disclosed by the end of August, and we will be working with the National Securities Regulatory Commission to analyze the environmental information disclosed by the listed companies and decide whether it is complete, accurate and integral.

Both annual report and semiannual report are known as regular reports. We are also pressing ahead with regulatory work of environmental information disclosure in the form of interim reports. For example, major incidents of listed companies such as severe administrative punishment or high legal risks due to environmental illegalities should be disclosed to shareholders and the general public. In recent years, some listed companies have failed or delayed to disclose their severe cases of environmental illegalities, which has violated the regulations. We are accelerating the statistical work concerning listed companies subject to major administrative punishment due to severe environmental illegalities.

Next, together with the National Securities Regulatory Commission, we shall make every effort to enhance the compliance, comprehensiveness, readability, integrality, and usefulness of information disclosure by listed companies to help shareholders and the general public to make informed judgement on environmental risks of these companies. Companies failing to promptly disclose their environmental information as required shall be urged and held accountable according to law and regulation, with necessary punitive measures taken.

Nanfang Metropolis Daily: According to some media reports, ozone has become another major pollutant affecting air quality after $PM_{2.5.}$ Is there any clear timetable for the control of ozone pollution?

Liu Youbin: Ozone is difficult to control due to its complicated mechanism of formation, and has not been effectively put under control even in developed countries. Related research and practice show that the control of ozone pollution requires synergetic cuts in NOx and VOCs emissions in certain proportion, hence differentiated control strategies are required based on pollution situation, industrial characteristics and economic and social development level.

China has included the prevention and control of ozone pollution into the agenda of atmospheric pollution prevention and control, and the percentage of days of good air quality is listed as a binding target in the outline of the 13th five-year plan, in which poor air quality due to ozone pollution is taken into consideration. *Action Plan on Air* published by the State Council clearly mandates controlling emission of ozone precursors such as VOCs and NOx. Currently, the MEP has reviewed and approved in principle the *Work Plan for Prevention and Control of Volatile Organic Matter Pollution during the 13th 5-Year Plan Period*, which is under discussion with other competent authorities. The Work Plan has clarified main objectives, priorities in control, major tasks and guarantee measures. Besides, Emission standards concerning VOCs have been developed and revised for 14 industries including petroleum refining, and charging policy of VOCs emission is published in collaboration with fiscal and pricing authorities to facilitate VOCs emission control. In terms of NOx control, NOx emission reduction remains a binding target in the 13th five-year plan.

From 2013, we have conducted ozone pollution monitoring and public disclosure of information in 74 key cities, which expanded to 338 cities of or above prefectural level in 2015. The next step shall be synergetic control of ozone and $PM_{2.5}$ with NOx and VOCs emission control as priorities by the following measures: first, strengthening the supervision and evaluation of binding target fulfillment of the 13th five-year plan; second, effectively cutting down precursors of ozone, i.e., NOx and VOCs, speeding up the pollution control of major industrial sectors; third, promulgating VOCs prevention and control policies and accelerating the standard formulation in chemical pesticide, paint, medicine and pharmaceuticals and unorganized volatility; fourth, conducting further research in the formation mechanism of ozone and the optimal proportion for synergetic reduction of NOx and VOCs in key zones. Through these efforts, it is expected that the growing trend of VOCs emissions and ozone pollution would be preliminarily curbed in the 13th five-year plan period.

The formation of ozone is related not only to industrial emissions but also to emissions from vehicles, home decoration, and apparel industries. From the health perspective, ozone pollution is relatively easy to prevent. The hazard of ozone pollution could be dramatically reduced with no significant outdoor exposure. The key is sensible understanding, rational response, and public participation. We hope that all members of society will adopt a green lifestyle to contribute to ozone pollution control.

CCTV: In the first half of the year, days of excellent air quality accounted for 74.1% on average in 338 cities of or above the prefectural level nationwide, a year-on-year decrease of 2.6 percentage points. The concentration of $PM_{2.5}$ in 13 cities in the Beijing–Tianjin–Hebei region areas registered at 72 $\mu g/m^3$, a year-on-year increase of 14.3%. How do you explain for such increase in $PM_{2.5}$ concentration?

Liu Youbin: Thank you for your question. First of all, I would like to provide you some statistics. The percentage of days of excellent air quality in 338 cities of or above prefectural level from January to July 2017 are, respectively, 60%, 69.8%, 83.1%, 83.6%, 70.5%, 77.8%, and 82.5%, and the concentration of $PM_{2.5}$ was, respectively, 117 $\mu g/m^3$, 62 $\mu g/m^3$, 48 $\mu g/m^3$, 40 $\mu g/m^3$, 38 $\mu g/m^3$, 29 $\mu g/m^3$, and 27 $\mu g/m^3$. Statistics shows that except for January and February, the percentage of excellent air quality remained above 70% in all months, and exceeded 80% during 3 months. The maximum level of PM2.5 was below 50 $\mu g/m^3$ except for January and February, with a monthly decline of, respectively, 17.20%, 4.8%, flat (unchanged), 3.3% and 6.9%.

Multiple incidents of severe air pollution occurred in January and February this year, which affected people's life and work. According to analysis by experts, the pollution was mainly caused by aggregate effect of unfavorable weather meteorological conditions amidst global warming, emissions from industries and vehicles and small-scale "scattered" coal burning for heating in winter. The frequent severe pollution in January and February caused the drop in percentage of days of excellent air quality in the first half of the year. The overall air quality has been continuously improving since March. Taking Beijing as an example, it is calculated that the severe pollution in January and February contributed 30% of $PM_{2.5}$ concentration in the first half of the year, while the average $PM_{2.5}$ concentration dropped by 25% on a year-on-year basis from March to June, hitting the record low for a succession of four months and gradually downsizing the bounce back. The air quality report of Beijing published yesterday showed that $PM_{2.5}$ concentration in July 2017 was 52 $\mu g/m^3$, a year-on-year decrease of 24.6%, again hitting the record low in history after March to June.

Practice shows that the arrangements for atmospheric pollution prevention and control in the *Action Plan on Air* have been effective. Meanwhile, we must be aware of the big gap between the current air environmental quality and public expectations, as illegal discharge of pollutants remains a challenging issue. Especially since the reinforced inspections for air pollution prevention and control were initiated, large numbers of enterprises in unauthorized locations, lacked the proper certificates and failed to meet emission standards causing pollution were found in various regions, reminding us that the high pressure for environmental law execution must continue without any slackening of efforts. We shall unswerving follow the uniform arrangements of the CCPCC and the State Council and make unrelenting efforts in defending our blue sky with confidence.

China News Service: The collection of environmental protection tax involves mobile factors such as water, air, noise, etc., hence is a challenging task. How is the execution of *Environmental Protection Tax Law* progressing now? How is the current progress on the determination of applicable amount of tax on local levels?

Bie Tao: Competent authorities are redoubling their efforts in drafting the regulations on the implementation of *Environmental Protection Tax Law*, while the notice on the implementation of Environmental Protection Tax Law has been co-published by the Ministry of Finance, the State Administration of Taxation and the MEP, requiring local authorities to prepare for tax collection and familiarize with the provisions of the *Environmental Protection Tax Law*. The conversion of sewerage charges to environmental protection tax requires the coordination, communication and information sharing mechanism between tax authorities and environmental protection authorities.

Environmental protection tax was formerly sewerage charge, which has been levied for over 30 years. According to the newly promulgated tax law, the first item on the agenda is to shift from charge to tax, i.e., the categories of current charges are shifted toward environmental protection tax. The basic working mechanism is as follows: enterprises shall report their quantities of pollutant discharge to tax authorities, who in turn verify the reported figures and correspondingly collect tax. Any doubt arising from this process may be reported to environmental protection authorities for review.

Concerning the adjustment and determination of applicable amount on local level, the tax is uniform and no major difference is allowed in principle; however, considering the specific environmental quality requirements and special need for industrial adjustment, the *Environmental Protection Tax Law* stipulates that the applicable amount of payable tax can be determined according to the table of tax amounts as provided in the Law on local level. For example, regions of less environmental capacity such as Beijing may set a higher payable amount of environmental protection tax. According to the principle of tax collection based on law, the procedures for determination and adjustment of tax amount on local level are as follows: proposal shall be submitted by provincial government to the standing committee of provincial people's congress for decision-making and to the Standing Committee of National People's Congress and the State Council for registration. Another basic principle is higher tax for larger quantity of pollutant discharge, and lower tax for less quantity of pollutant discharge, i.e., the tax shall be cut by 50% if the pollutant discharge is 50% lower than the standard, and by 75% if the discharge is 30% lower than the standard.

As the basis for environmental protection tax, the quantity of discharged water and air pollutants can be calculated by one of the following four methods, where are to be applied based on different orders of priority. First, monitoring data are automatically acquired by monitoring equipment installed in major pollutant-discharging enterprises, which are deemed reliable unless tampered with, and the operation and maintenance are dependable. Such data are on the top of the priority list. In the case where automatic monitoring is not feasible, manual monitoring is also acceptable and the quantity of pollutant discharge shall be determined based on monitoring data. The third method is a combination of two methods: discharge coefficient method and material balance method. In order to reduce the cost of tax collection to a reasonable level, for enterprises covering large areas or with large quantities of pollutant discharge, or where discharged pollutants are highly varied, hence unable to monitor, discharge coefficient method and material balance method shall be adopted by the MEP to calculate the quantity of pollutant discharge. To provide a simple example, suppose the sulfur content of coal used by a certain power station is 5% and the total consumption of coal is 5 million tons, the quantity of sulfur pollutant discharged shall be the total coal consumption multiplied by 5%. The material balance method refers to the calculation based on basic consumption of raw materials, products manufactured and waste produced in the same industry, based on the principle of material conservation.

Currently, the MEP and the State Administration of Taxation have signed a memorandum of understanding concerning the collection of environmental protection tax. Next, the environmental protection authorities will continue to work closely with fiscal and tax authorities to make all necessary preparations for the collection of environmental protection tax to guarantee a smooth start on January 1st, 2018.

Beijing News: According to some recent media report, with increasing pressure of environmental protection, the "one-cure-for-all" policy is being applied by some local governments to restrain livestock and poultry breeding, relocating and closing down various types of animal farms. Is this really happening? What measures are being taken to cope with this situation?

Liu Youbin: China is a major producer of livestock and poultry products, and environmental problems arising from the development of livestock and poultry breeding have become evident. According to the statistics from the Ministry of Agriculture, approximately 3.8 billion tons of manure is produced by livestock and poultry annually, of which 40% are not efficiently utilized or treated for biosafety, which has incurred severe environmental impact and thus becomes a major environmental issue in rural regions.

In order to effectively tackle the environmental problems arising from livestock and poultry breeding and promote its sustainable development, the state has promulgated well-structured laws and regulations for the control and prevention of pollution from livestock and poultry breeding, with comprehensive provisions on the planning, site selection and construction of large-scale breeding farms, pollution prevention and control in high-density areas of free-range breeding, zoning of livestock and poultry forbidden areas and corresponding compensation, etc., providing legal guarantee and basis for the prevention and control of pollution from livestock and poultry breeding.

Going forward, the MEP shall perform its administrative duties according to law, guide and urge the local authorities to prevent and control pollution by livestock and poultry breeding according to law, take rectifying measures on enterprises failing to meet environmental standards to urge their upgrading and strengthen support for law-abiding enterprises to safeguard their legitimate interest.

Jiemian News: Recently the Intermediate People's Court of Maoming, Guangdong Province ruled that environmental protection organizations such as Chongqing Liangjiang had no right for public interest litigation for marine environment. This ruling was called into question by the public and some experts claim that such ruling was not well grounded. What is your opinion? Is there any plan to lower the threshold of plaintiff qualification for public interest litigation concerning environmental issues?

Bie Tao: Thank you for your interest in this issue and this case. We have also noted that it is an interesting case. The plaintiff is an active environmental organization located in Chongqing, and I appreciate their engagement in supervising the environmental protection. As what is happening in Maoming is a judicial case, it would be inappropriate for an administrative authority like us to pass any judgement.

Based on my understanding, the ruling of the court is not groundless. According to the *Marine Environment Protection Law*, in the case of damage to marine ecological environment causing severe loss for the country, authority exercising the right of supervision and administration of marine environment according to the law may initiate litigation for damage compensation on behalf of the country. As is interpreted by the court, it is marine supervision and administration authorities that are entitled to the litigation right stipulated in the *Marine Environment Protection Law*, and NGOs are not explicitly included.

Currently, different opinions are expressed by media and law experts, which I think is understandable. *The Environmental Protection Law* revised in 2014 explicitly stipulates that social organizations meeting certain criteria have the right to initiate public interest litigation according to law. Recently the *Civil Procedure Law* and the *Criminal Procedure Law* have been revised by the Standing Committee of the National People's Congress, stipulating that environmental public interest litigation may be initiated by environmental protection organizations or procuratorates. Yet it remains unclear whether public interest litigation by NGOs and litigation initiated by authorities are mutually exclusive or not. And we have encountered issues concerning the coordination between litigation for eco-environmental damage compensation by government and public interest litigation by NGO in our pilot reform of ecological environmental damage compensation system. Some good practices are already enabled at local level. In some places, public inter-test litigations against enterprises causing pollution initiated by NGO which have not been adjudicated by the court are consolidated with government-initiated litigation for trial. I think this is a good practice. With the consolidation of cases, the evidences held by the government and those by NGO can complement each other and be used together to prove the illegal behavior of the enterprise and hold it accountable. Whether the case is brought to court by the government or by NGO will not put law-breaching enterprises in any favorable situation. Leveraging on the respective advantages of the government and the NGO will help to curb the discharge or emission of pollutants.

Concerning the "one-cure-for-all" issue you brought up, I would make one final clarification with emphasis that the so-called one-cure-for-all is never a requirement by the MEP. Thank you.

Liu Youbin: This is the end of today's press conference. Thank you.

Regular Press Conference (September)

September 27, 2017

The regular press conference of MEP in September was hosted on the morning of September 27, where Mr. Cui Shuhong, Director General of the Department of Environmental Impact Assessment of MEP, briefed on the environmental impact assessment work, and Mr. Liu Youbin, Counsel (Director-General level) of the Department of Communications and Education, reported on recent progress in prioritized areas in environmental protection. The two officials jointly responded correspondents on questions of interest, including whether environmental protection would affect economic development.

Ministry of Environmental Protection of the People's Republic of China,
2017 Press Conference Records of Ministry of Environmental Protection, the People's Republic of China, https://doi.org/10.1007/978-981-13-7330-5_10

MEP September Press Conference (1)

MEP September Press Conference (2)

Liu Youbin (Moderator): Dear friends from the media, good morning! Welcome to the September session of regular press conference of MEP.

Environmental impact assessment is an important institutional system in environmental protection. In recent years, much has been done in environmental impact assessment reforms with evident progress. Mr. Cui Shuhong, Director General of the Department of Environmental Impact Assessment is here with us. Later he will brief you on the situation of environmental impact assessment in China and answer your questions of interest.

Now I would like to report briefly on recent progress in several prioritized areas in environmental protection.

1. Evident progress in pilot reform moving toward vertical administration system in environmental protection

On September 14, 2016, *Guiding Opinions on Pilot Work of Vertical Administration System in Monitoring, Inspection and Law Execution by Environmental Protection Authorities below Provincial Level* (hereinafter referred to as *Guiding Opinions*) was published by the General Offices of the CPC Central Committee and the State Council in preparation for the launching of the vertical administration reform in environmental protection. MEP and the State Commission Office of Public Sector Reform have conscientiously implemented the work in a down-to-earth manner, providing specified guidance with respect to different situations. Currently, pilot provinces (municipalities) where plans have been made include Hebei, Chongqing, Jiangsu, Shandong, Hubei, Qinghai, Shanghai, Fujian, etc., where the vertical administration system of environmental protection is established effectively. Among them, institutional readjustment is by and large concluded in Hebei and Chongqing; efforts are being made to enable the reform in Jiangsu and Shandong, Hubei, Qinghai, Shanghai, and Fujian are in the process of registration; while the implementation plan for vertical administration reform is being reviewed on provincial (municipal) level in Shaanxi, Jiangxi, Tianjin and Guangdong. Evident progress is made in pilot work of the vertical administration reform.

As mandated by the Guiding Opinions, the system of appointment and dismissal of leaders determined mainly by municipal (prefectural) environmental protection bureau is replaced by dual administration with the provincial environmental department (bureau) playing the leading role; meanwhile, county (district) level environmental protection bureau is under direct supervision by municipal (prefectural) environmental protection bureau, and the appointment and dismissal of leaders is determined on the municipal (prefectural) level. The basic approach for environmental supervision system reform has been shaped in the pilot provinces, i.e., a number of divisions or offices are established in the provincial environmental protection department; the latter dispatches environmental supervision agencies across cities (counties) or in each city as permanent "environmental protection inspection group of provincial CPC commission and government." Current municipal (prefectural) environmental monitoring agencies and staff are under the direct supervision by provincial environmental protection department and are responsible for conducting ecological environmental quality monitoring, survey, and assessment of their dispatched locality in an independent and objective manner. After the reform, "supervision of enterprises" has become the focus for environmental execution in pilot provinces, and the law enforcement on the county level is subject to uniform supervision and command of the municipal level for cross execution and joint execution.

As mandated by President Xi Jinping, the vertical administration reform in environmental protection should focus on tackling the "four salient issues" prevailing in the sectioned administration system of environmental protection, namely: difficulty in exercising supervision over local government and other competent authorities, intervention on environmental inspection, monitoring and law execution by local protectionism, failure in finding solutions to cross-regional and inter-basin environmental issue in an coordinated way and ineffectiveness in standardizing and strengthening capacity building of regional environmental protection authorities. The pilot reforms have successfully explored for an effective path in tackling the "four salient issues," with the goals by and large identified for the reform of local environmental protection administration system, i.e., combination of zones and sections, clear accountability and functions, well-defined power and responsibilities, adequate logistic support, authority and efficiency. The experience, models, and practices formed by the pilot provinces are replicable and reproducible, hence building a solid foundation for expansion of the vertical administration reform in environmental protection from successful pilot projects to massive implementation.

As required by the *Guiding Opinions*, the vertical administration reform in environmental protection in other provinces shall continue based on the principle of "registration and launching of each project that all conditions are mature," aiming at completing the systematic reform in all environmental protection authorities under the provincial level by the end of June, 2018, while perfecting the auxiliary measures and institutional mechanism so as to accomplish the mission of vertical administration reform in monitoring, inspection and law execution by Environmental Protection Authorities during the 13th five-year plan period, ensuring that all environmental protection departments under the provincial level shall function efficiently based on the new system by 2020.

2. Approval for ecological protection red line completed in 10 provinces and municipalities

On February 7, 2017, the General Offices of CPC Central Committee and the State Council published *Opinions on Delineating and Strictly Enforcing Ecological Conservation Red Lines*. As arranged and required by this document, MEP, in collaboration with the National Development and Reform Commission and with joint efforts with other competent authorities and local government to press ahead with the delineation and observation of the ecological conservation red line.

First, a coordination mechanism is established for all tasks to be handled in a coordinated way. MEP and NDRC have initiated the establishment of an inter-ministerial coordination steering group for ecological protection red line which consists of 12 institutional members, hosting expert panel meetings under the inter-ministerial coordination steering group to strengthen technical guidance for ecological protection red line.

Second, top-layer design is strengthened with the publication of guidance documents. MEP joins other competent authorities in publishing *Work Plan for Implementing the Opinions on Delineating and Strictly Enforcing the Ecological Conservation Red Line*, *Guidelines for Delineating the Ecological Conservation Red Line*, *Technical Review Procedures of the Plan for Delineating the Ecological*

Conservation Red Line, Work Plan for Supervision of Delineating and Strictly Enforcing the Ecological Conservation Red Line, providing technical guidance and fundamental work guidelines for delineating and strictly enforcing the ecological protection red line.

Third, liaison between MEP and regions has been established to provide instructions and guidance in delineating the ecological conservation red line. We have jointly hosted numerous meetings with NDRC to coordinate the defining of red lines in key areas including Beijing–Tianjin–Hebei Region and provinces (municipalities) along the Yangtze River Economic Belt, giving instructions and facilitating cross-regional coordination. Up to now, the review and approval for ecological conservation red lines defined in 10 provinces (autonomous regions and municipalities), namely Ningxia, Shanghai, Beijing, Tianjin, Chongqing, Yunnan, Jiangxi, Hubei, Anhui, and Sichuan, are completed.

Next, MEP will collaborate with NDRC and other competent authorities to carry forward defining and observing the ecological conservation red line, publishing guidelines including Opinions and Suggestions for Spatial Pattern and Layout of Ecological Conservation Red Line for All Provinces (Autonomous Regions, Municipalities), etc., while strengthening technical guidance to provinces (autonomous regions, municipalities). By the end of the year, Beijing–Tianjin–Hebei Region and provinces (municipalities) along the Yangtze River Economic Belt shall finish defining their red lines with our support and guidance, and the national supervision platform for ecological conservation red line shall be launched into trial operation as scheduled. By the end of 2018, all other provinces (autonomous regions, municipalities) shall also finish defining their red lines.

3. Research projects undertaken by 28 expert teams in "2+26" cities

It is clearly stated in *Research Program on Causes and Control of Heavy Air Pollution* that "one-on-one" responsibility system for follow-up studies shall be put in place in "2+26" cities, appointing 28 expert teams and dispatching one team to each city to provide on-site guidance, obtaining first-hand data of prevention and control and proposing comprehensive solutions for air pollution with city-specific strategies.

Each work group, constituted by one initiator and three to six participating agencies, consists of both national-level experts and local researchers, and the size is adjustable based on the size of the city and the actual situation of air pollution in autumn and winter, ensuring sufficient workforce for follow-up studies. Meanwhile, three technical expert panels, respectively, specializing in pollution source tracing and cause analysis, compilation and control of list of pollution sources, comprehensive decision-making and supervision platform construction, providing fundamental support with science and technology and ensuring the normativity of follow-up studies as well as the consistency and comparability of research results.

The follow-up studies will enable us to acquire reliable and detailed first-hand data of local air pollution, which in turn provide support for decision-making in atmospheric environmental management; second, such studies can be closely integrated with air pollution prevention and control efforts on local level, helping local authorities to tackle bottleneck issues and narrowing down the gap between research and application; third, training for local personnel in air pollution control and prevention is provided through the follow-up studies, which contributes to capacity building on local level; fourth, support in science and technology is provided for current priorities on the agenda of environmental protection, such as action of comprehensive atmospheric pollution control in Beijing–Tianjin–Hebei Region and its periphery areas in autumn and winter.

Up to now, 28 research teams are dispatched to local level to take up their work.

Now I would like to invite Director General Mr. Cui Shuhong, to brief you on the progress of ongoing work.

Mr. Cui Shuhong, Director General of the Department of Environmental Impact Assessment, MEP

Cui Shuhong: Dear friends from the media, good morning.

First of all, thank you for your keen interest in and support for environmental impact assessment over time. It is my great pleasure to meet you and answer your questions.

Environmental impact assessment (EIA) is an important system of prevention, which is at the very source of environmental protection. Since the 18th CPC National Congress, we have been conscientiously implementing the arrangements and requirements of the CPC Central Committee and the State Council in terms of changing the governmental functions, streamlining administration and delegating power to lower levels as well as institutional reform in ecological civilization. We have adopted a problem-based approach and invested much effort in changing the means and methods of environmental administration centering upon environmental quality improvement, developing and publishing *Implementation Plan for*

Environmental Impact Assessment Reform during the 13th 5-Year Plan Period. Positive progress has been made in EIA reform.

First, legal system establishment has seen major breakthroughs. The newly revised *Environmental Impact Assessment Law* came into effect on September 1st, 2016, marking another breakthrough in environmental protection legislation since the revision and promulgation of *Environmental Protection Law*. Recently, the State Council has published the newly revised *Administrative Regulations on Environmental Protection of Construction Projects*, which is to be implemented since October 1st this year.

Second, EIA procedures continue to be simplified, with the preapproval for water and soil conservation and industrial preview cancelled, and the EIA itself is no longer a precondition for approval by other departments, with the EIA approval switching from a "serial" to "parallel" procedure, an integrated approval approach. Registration has replaced review and approval procedures for items on the registration form, which account for approximately 50% of all cases for environmental impact assessment approval, which has significantly cut the administration cost for grassroots environmental protection authorities as well as burden on enterprises. Permit for trial production of construction projects and administrative permit for environmental protection approval on completion are both abolished and replaced by self-examination and approval by constructors.

Third, the classified and categorized administration is further optimized. The MEP has delegated power of EIA approval for construction projects for 57 items in two batches. Since 2015, 28 provincial environmental protection authorities published new EIA catalogues and revised *Catalogue of Classified Administration of Environmental Impact Assessment for Construction Projects*, reflecting scientifically based adjustments and improvement in the EIA categorization of projects.

Recently we focus on pressing ahead with reforms in three areas:

First, in order to tackle the difficulty in implementing the strategies and plans, we have defined the restraints on macro level and strengthened the binding power of the ecological protection red line, environmental quality bottom line, maximum limit of resource utilization and negative list for environmental access, thus highlighting space access and environmental administration. Definition of these lines, limits, and lists is carried out as pilot program in Lianyungang, Chengde, Ordos, and Jinan, and the technical guide for defining the lines, limits, and lists is preliminarily compiled.

Second, in order to address the unclear boundaries and targets of project EIA, rules are defined in the project EIA to facilitate restructuring of the technical guide system to reach the desired objective of EIA. We have completed the revision of *Outline of Technical Guidance of Environmental Impact Assessment for Construction Projects* and are currently revising and completing the guidelines for technical guide for pollution source intensity calculation and technical guide of EIA for thermal power, pulp making, paper-making, steel and cements plants upon soliciting public opinions. The formulation and revision of technical guide for EIA of surface water environment, atmospheric environment, soil environment, and environmental risk are progressing with intensified efforts as well.

Third, to address deficiencies in interim and ex post regulation, supervision on implementation is highlighted together with institutional and systematic reforms and further clarification of the primary responsibility of environmental protection of constructors. Spot check system with random choosing of inspectors and those being inspected with a clearly defined objective and information disclosure has been established with stronger punitive measures adopted on law-breaching practices. EIA informatization is accelerated, with "network + database + platform" as the priority, i.e., national EIA review and approval information network, EIA underlying database and intelligent EIA regulation platform, and the implementation plan for interim and ex post regulation reform is being studied and developed.

Generally speaking, the EIA reform is now progressing smoothly and effectively. The background information is in the printed handouts. Thank you.

Liu Youbin: Now I would like to take your questions.

China Youth Daily: There is a lot of badmouthing about the environmental impact assessment system on the internet, claiming that such system is at the point of fading out and quitting the stage. What is your opinion about this?

Cui Shuhong: Thank you for your question. This is a good question and we appreciate constructive input for EIA system reform; however, I must tell you that such claim on the internet is not fact-based and is therefore incorrect. As an important system of environmental protection administration, EIA system will be strengthened rather than weakened as the reform progresses. We know that EIA is an international practice to prevent environmental pollution and ecological destruction from the source and has been implemented in China for almost 40 years. Currently, China has put in place a well-structured institutional system for EIA, which plays a significant role in preventing environmental pollution and ecological destruction. Meanwhile, it has effectively promoted the healthy economic and social development while safeguarding the environmental rights and interests of the people.

Environmental issues have increasingly become a salient problem with the social and economic development of China; in particular, with new requirements for the shift in governmental functions, some weaknesses are observed in the method and model for EIA management, such as excessive items for review, difficulty in implementing of the planning environmental impact assessment, unclear target of project EIA as well as the frequent occurrence of illegal practices such as construction before approval and procrastination in inspection.

To resolve such problems, we need to be problem-oriented, highlighting the delegation of power, administration and service, and focus on defining the limits, setting the rules and supervising the implementation with environmental quality improvement at the core. This represents the main direction of the current EIA reform. Through defining the limits, setting the rules and supervising the implementation, we have redoubled our efforts in EIA reform so as to meet the desired goals. In the first year of the 13th five year-plan period, we have developed and published the *Implementation Plan for Environmental Impact Assessment Reform during the 13th five Year Plan Period*, covering all areas of EIA including strategy, planning and project and the entire EIA process consisting of 46 reform tasks. This Plan is a top-level design for EIA reform as well as the blueprint and roadmap of reform. The main areas of reform as defined in the plan include the following: first, strengthen implementation of EIA in project planning through spatial management and definition of limits; second, set the rules through revision of guidelines for better pertinence of project EIA, which used to be a mélange of everything due to unclear boundaries, and the revision of guidelines is expected to bring EIA back to its original purpose; third, strengthen interim and ex post regulation through inspection on the implementation; fourth, guarantee information transparency and public participation through standard procedures.

Meanwhile, major actions of reform are clarified in the Plan:

First is to set the limits. Through strengthening the ecological protection red line, environmental quality bottom line, maximum limit of resource utilization and negative list for environmental access, which are collectively known as the lines, limits, and lists, we put into place rigorous constraints to ensure the implementation of EIA on strategic and planning levels.

Second is to establish and strengthen an interactive and connected mechanism between project EIA approval and planning EIA, environmental management of existing projects and regional environmental quality to serve environmental protection as a priority task, of which the key is to improve environmental quality.

Third is to conscientiously safeguard people's environmental rights and interests through synergy of three elements, i.e., law-abiding business operations by enterprises, enhancing information transparency and public participation, and strengthening guidance of public sentiments.

Fourth is to intensify efforts in the four-level national EIA review and approval information network. Environmental protection authorities on national, provincial, municipal and county levels are required to report review and approval information to a platform through the four-level network for analysis enabled by big data, hence fundamentally transforming the model of interim and ex post EIA management.

I believe that the preventive function of the EIA system will be further strengthened through these measures. Thank you.

China News Service: The term "the line, limit, and list" has appeared in the printed handouts. Could you please elaborate on what is to be done exactly on the agenda of "the line, limit, and list"? How would it contribute to eco-environmental protection in China?

Cui Shuhong: Thank you for your question. Just now I mentioned that the cardinal line in EIA reform is defining the limits, setting the rules and supervising the implementation. The key measure in defining the limits is to define the line, limit, and list. President Xi Jinping emphasized during the 41st study session of the Political Bureau of the CPC Central Committee that we must redouble our efforts in setting the three red lines, namely baseline of ecological function protection, bottom line of environmental quality and safety and maximum limit of natural resource utilization, facilitate the evolvement of a green development path and lifestyle and accelerate the construction of a scientifically based, moderate and orderly layout system of land resources. Defining the line, limit and list is the implementation of President Xi Jinping's important instructions.

The contents of line, limit, and list include the red line of ecological protection, the bottom line of environmental quality, the maximum limit of resources utilization and the list of environmental access. My colleague Cheng Lifeng, Director General of the Department of Nature and Ecology conservation, MEP, has just briefed you on this point. The core is to ensure that the red line of ecological protection "may expand but will never recede" by defining this red line, setting spatial limits for economic development layout. This enables us to determine what projects are not allowed within the red line, thereby strengthening the control of spatial layout.

As for the bottom line of environmental quality, the core task for environmental protection is to improve the environmental quality. Setting the bottom line of environmental quality helps to define a mandatory redline for enhancing the environmental quality, which must improve rather than deteriorate. This is our bottom line.

The bottom line of environmental quality is defined to set boundaries on the scale of economic development and the intensity of development of a region. For example, China has set rather strict limits in SO_2, NOx and dust emissions of thermal power plants as compared to the international standards, and the emission per plant is quite low. However, the aggregate emissions may exceed the environmental capacity if numerous plants are built in the same region. Despite the excellent environmental performance in emission control of each plant, the total emissions from multiple plants may exceed the limit of environmental capacity, resulting in environmental quality deterioration of the region. By defining the bottom line of environmental quality, the development scale and intensity shall be capped.

Concerning the maximum limit of resources utilization, we all know about the dual property of resources, namely economic property and environmental property. The media of all environmental factors are resources, such as water, land, forest, grassland, etc., and the abusive exploitation of resources is one of the key factors for environmental deterioration. Therefore, setting the maximum limit of resources utilization is based on the understanding that resources are priced in terms of ecological value, and the exploitation of resources should ensure the preservation or appreciation of their value, especially for resources with environmental carrying capacity. Besides, the economic property of resources means that exploitation and utilization are inevitable; based on the utilization rate of resources and considerations for the preservation and appreciation of resource value, we can define the maximum limit of resources utilization in a region, which is a ceiling of resources exploitation intensity.

Based on the red line of ecological protection, bottom line of environmental quality and maximum limit of resources utilization, we would consider multiple factors in spatial layout, intensity and scale of regional development, and determine the prerequisites for environmental access for industries, which is what we call the negative list. Projects which do not meet the environmental access prerequisites are not allowed to be constructed in this region. This will fundamentally optimize the regional industrial layout and solve the environmental problems incurred by abusive exploitation and utilization of resources and excessive development in the region.

We have carried out pilot work in setting the line, limit and list, with the technical plan preliminarily taking shape and we are currently soliciting feedback from various regions as well as input from experts. The technical guide of defining the line, limit and list will soon be published based on expert suggestions, opinions from local level and experience from pilot work to provide guidance and instructions for carrying out the work in all regions.

It can be safely stated that setting the line, limit and list constitutes an important element in the reform of environmental management as well as in the reform of EIA, which will contribute enormously to China's environmental protection.

The second question is about the roles played by EIA. Here I have some statistics to share. It is fair to say that EIA plays a significant role in controlling and optimizing the development intensity, spatial layout and development scale. For instance, conditions for environmental access has become more stringent; during the 12th five-year plan period, over 1,800 projects which did not meet the environmental access prerequisites were rejected for review and approval by environmental protection authorities on ministerial and provincial levels. Through EIA for four hydropower station projects in Western China, 25 cascade layouts were cut and over 1,170 km of natural rivers were preserved, increasing the natural river preservation rate from 37.4% to 71.3%. For example, in the upstream of Jinshajiang River, the original plan was 13 cascades, of which 5 were eliminated through planning EIA, and the rest 8 cascades were permitted by environmental conditions, significantly reducing the development intensity. Through planning EIA for coastal port, 34 natural reserves were avoided in the development, 173 km of planned coastal line was cancelled, and land reclamation was reduced by 224 km^2. Numerous examples show that project EIA significantly cuts the total pollutant emissions of a region. In the coal-fired power resource development in the periphery areas of Beijing–Tianjin–Hebei Region, the amount of coal utilization was reduced by more than 29 million tons through planning EIA consultation. All these will effectively facilitate the regional environmental quality improvement.

Jiemian News Service: According to some recent media reports, an environmental protection organization in Tianjin applied for disclosure of EIA results. And only 2 out of 29 steel plants provided the information as requested. What measures would be taken by the MEP to tackle the barrier in EIA information disclosure by enterprises or governmental authorities as requested by the public or by environmental protection organizations?

Cui Shuhong: Thank you for your question. It should be pointed out that information transparency is the prerequisite and condition for ensuring the right to know, the right of participation and the right of supervision of the public. The MEP attaches great importance to information transparency, following the principle of "information disclosure as a normal practice and nondisclosure in exceptional cases." Tremendous efforts are made to facilitate the environmental information transparency, with numerous documents successively published including *Measures of Environmental Information Disclosure (Trial), Guide for Information Disclosure of the Ministry of Environmental Protection, Catalogue of Information Disclosure of the Ministry of Environmental Protection (Batch 1)* and *Measures of Environmental Information Disclosure by Enterprises and Public Institutions*. In terms of environmental impact assessment, we have published *Provisional Measures of Public Participation in Environmental Impact Assessment*, which is currently being revised, and printed *Guide for Governmental Information Disclosure of Environmental Impact Assessment of Construction Projects*. From the regulatory perspective, requirements for information transparency have covered multiple areas, which is explicitly stipulated in the *Environmental Protection Law* as you can see. In 2016, the MEP took the initiative to publish 16,432 articles disclosing governmental information of various types, and our official website was visited for a total of 710 million times by 91 million visitors, with an average of nearly 7.6 million visitors every month. All these played a very positive role in facilitating information transparency.

However, what we have achieved in terms of information transparency by now is not impeccable. The first challenge is the reluctance in information disclosure; the second is irregularities in information disclosure and the third is failure in responding to public inquiries in a timely and thorough manner after disclosure. My opinion on information transparency is that each stakeholder should conscientiously fulfill their respective obligations in strict compliance with relevant rules and regulations and on a legal basis. Those who fail to perform the responsibility of information disclosure should be urged by their governing authorities to disclose the information as required; while irregularities in information disclosure by local environmental protection authorities would be notified by us for rectification.

Besides, public supervision also promotes information transparency, including information transparency of enterprises. The MEP has recently taken some measures for information transparency of enterprises and public institutions. Taking waste incinerators as an example, the MEP currently requires all waste-to-power plants to install automatic pollutant monitoring equipment, set up a big screen in front of the plant to disclose environmental and emission information to public and connect the automatic monitoring equipment with local environmental protection authorities' network.

National Business Daily: The MEP has recently published *Guide for Environmental Impact Assessment of Hazardous Wastes in Construction Projects*. Could you please provide some information on the current situation and weaknesses in EIA of hazardous wastes in China? What problems are expected to be solved with the implementation of the Guide?

Cui Shuhong: As a priority as well as the biggest challenge in environmental protection management at the moment, the management of hazardous waste is a key requirement in the *Action Plan on Soil* and the *Plan of Ecological Environmental Protection During the 13th Five Year Plan Period*. As an important element in China's legal framework of environmental protection, EIA has always placed solid waste among the top of its agenda. The recently formulated *Outline of Technical Guidelines of Environmental Impact Assessment of Construction Projects* as I have just mentioned has stipulated requirements for EIA of hazardous waste. However, we did notice that as no specific normative technical guide for EIA of hazardous waste has been made in China, some contents in the EIA document concerning solid waste are not systematic, standardized or clear, with weaknesses in the interface between EIA and regulation during operation, therefore failing to play an effective supporting role for strengthening environmental regulation of hazardous waste from the source.

To tackle these problems. The MEP recently published *Guide for Environmental Impact Assessment of Hazardous Waste in Construction Projects* (hereinafter referred to as the *Guide*), through which we aim at solving some problems. First is the innocuity, reduction and recycling of hazardous waste. According to the *Guide*, the entire process of generation, collection, storage, transport, utilization and treatment of hazardous waste in construction projects should be assessed. Meanwhile, technical requirements for such assessment which used to be weak or unspecific is now clarified in the *Guide*, expecting that normative standards should be set in engineering analysis, environmental impact analysis, technical and economic viability of pollution prevention and control measures, environmental risk assessment, environmental management requirements, specific conclusions and recommendations, etc. Requirements for hazardous waste are also specified in environmental impact report and list under the corresponding theme.

In summary, it is fair to say that the publication of this *Guide* is an important measure to standardize and strengthen hazardous waste management and to narrow the gap in environmental management. Thank you.

Beijing Youth Daily: Recently an EIA professional described the EIA report he wrote on Zhihu,[1] the Quora of China that "many contents are imagined and many data are fake," "my daily work is copying and pasting reports." How does your Ministry view such opinion expressed by an EIA professional that the EI report is not factual-based? Some people claim on the internet that some small EIA institutions are solely profit-driven and their behaviors are severely disrupting the market. How to ensure the impartiality and accuracy of third-party EIA institutions?

Cui Shuhong: Thank you for your question. This is one of the problems to be tackled in EIA reform. Most of EIA professionals are doing a good job, but in some places, there are still frauds. There are clear stipulations from both the legal perspective and the regulatory perspective about frauds in EIA. All fraudulent behaviors are subject to severe punishment according to law.

---

[1]Zhihu: A Chinese portal for Q&A to share knowledge, experience and opinions.

A series of measures are taken to crack down frauds by certain EIA institutions and individuals, including leasing or lending EIA qualification. First, we scrutinize the quality of EIA report and list, requiring strict examination during the reception and review of environmental report and list by environmental protection authorities of all levels. Reports and lists with fabricated contents and inaccurate prediction for EIA conclusion shall not be approved, and institutions and individuals accountable for such reports of poor quality or false information shall be severely punished.

Besides, we have strengthened the guidance for environmental protection authorities on local level. I have already mentioned the four-level EIA information report system which channels the information of projects reviewed in various regions onto one platform, which is known as intelligent EIA supervision platform. Enabled by big data, we verify the EIA review quality of these projects as well as consistency of the quality of report and list, spotting problems for reinforced guidance and supervision.

More severe punitive measures are put in place for illegal practices in EIA. Institutions that intentionally practice frauds either have their qualifications revoked or are ordered to rectify their behaviors within a fixed period of time. Moreover, the EIA credibility system is reinforced by setting a specific column on the official website of the MEP to blacklist the institutions and individuals guilty of fraudulent or deceitful behaviors.

To respond to your second question, we are highly aware of the misconduct of some small EIA institutions during the reform, which have disrupted the market. As is known, institutions producing EIA reports and forms must be qualified. Without qualifications, no institution is allowed to operate on EIA market. Yet malicious bidding by some enterprises on the EIA market has disrupted the market order, so our management on EIA institutions has always been strict.

First, we have launched a series of special inspections and spot checks. Last year we organized a special inspection on market behaviors of EIA institutions, and this year's inspection is currently in progress with another round scheduled for the second half of this year to rectify market behaviors.

Second, *Administrative Measures for Qualifications of Environmental Impact Assessment of Construction Projects* and *Regulations on the Environmental Protection of Construction Projects* revised in 2015 have delegated the power of supervision and punishment to local environmental protection authorities, which helps to boost synergy for stricter management of EIA institutions.

Third, the accountability and investigation mechanism must be strengthened, including holding the institutions and individuals responsible for poorly produced EIA reports accountable. Information transparency should also be promoted for public supervision. I have just mentioned the building of credibility system. Since 2015, 362 EIA institutions were investigated and punished for their illegal or incompliance behaviors by environmental protection authorities on all levels, of which 14 institutions guilty of the severest cases of misconduct had their EIA qualifications suspended or revoked. In December 2016, the MEP created a system which so far has registered over 730 entries concerning credibility, including 376 entries on fraudulent conduct by EIA institutions and 329 entries on misconduct by

EIA engineers. As the compilation of EIA report and form is a market behavior, project owners may make inquiries in our credibility system so as to avoid institutions or individuals with a record of misconduct. Thank you.

China National Radio: Recently, two environmental protection organizations: Institute of Public and Environmental Affairs (IPE) and the Natural Resources Defense Council (NRDC), published the evaluation results for Pollution Information Transparency Index (PITI) of 120 cities in 2016–2017. What achievements have been made in China in enabling pollution information transparency? What are the problems? What measures are to be taken next?

Liu Youbin: We have noticed these reports published by relevant social organizations. A chapter specifically on information transparency is included in the newly revised *Environmental Protection Law*, which fully reflects the importance attached by the central government on environmental information disclosure. The MEP also gives much attention to information transparency, considering environmental information disclosure as an important pivot in strengthening environmental protection according to law. We appreciate the active involvement of social organizations in promoting environmental information transparency.

Pollution information transparency and corporate environmental information disclosure are two pivotal aspects of environmental information transparency. In terms of pollution information transparency, the MEP has issued the notice on strengthening pollution information transparency as early as in 2013. At the moment, all provincial, prefectural and municipal environmental protection authorities have included the column of pollution information which is accessible by public, disclosing environmental regulatory information including pollutant monitoring on their governmental portal websites. Meanwhile, the MEP updates the public on the implementation of supporting measures for *Environmental Protection Law* on a monthly basis.

In terms of corporate environmental information disclosure, the MEP published *Measures of Information Disclosure by Enterprises and Public Institutions* (hereinafter referred to as *Measures*) in 2014. Since its publication and implementation, nearly half of the prefectural level cities have disclosed the list of main pollutant-discharging entities in the locality as required; however, problems still prevail as some provinces and more than half of the prefectural level cities have not yet published their lists, information disclosure is progressing rather slowly and some disclosed environmental information is difficult to inquire, etc.

To make environmental information disclosure more effective, the MEP has further revised the *Measures*. First, in order to enable easier environmental information disclosure, the requirements for corporate environmental information disclosure are simplified and clarified in strict compliance with the *Environmental Protection Law*.

Second, to make inquiries and supervision by public more convenient, a uniform corporate environmental information disclosure platform shall be established on the national level, on which main pollutant-discharging entities shall take the initiative to disclose their information and be responsible for the truthfulness and accuracy of the published information.

Third, to urge environmental protection authorities to supervise enterprises on their fulfillment of information disclosure obligation, requirements including the frequency of dynamic pollutant discharge information disclosure by enterprises should be further clarified according to the current administrative system.

The revised draft of the *Measures* has been circulated among all provincial environmental protection departments and other competent authorities in China to collect their feedback for further revision from May to July this year, and has been published on the website of Legislative Affairs Office of the State Council and the MEP official website for open solicitation of opinions from the public. Next, we are going to publish and implement the revised version of the *Measures* after some further changes and improvements to facilitate the environmental information disclosure. Thank you.

Lianhe Zaobao (Singapore): The number of days with clear sky increases whenever major events are held in Beijing. We have experienced "the Military Parade Blue" and "APEC Blue." As the 19th National Congress of the CPC is to be held soon, what measures would be taken to guarantee the air quality? According to some media reports, environmental problems related to air quality prevail in half of the 40,000 enterprises in Beijing–Tianjin–Hebei Region and its periphery areas inspected by the end of August, and the task of atmospheric environmental governance remains a daunting task. Will stricter measures be implemented?

Liu Youbin: Thank you for your question. The overall environmental situation in China allows for no blind optimism and the control of air pollution remains in a key phase with many daunting tasks and challenges, which is especially visible in the gap between cities.

We are fully aware of the severity of this challenge and have never slackened our efforts. And it is heartening to see positive momentum in environmental quality improvement in some regions as the entire society attaches much importance on environmental protection, and the speed of such improvement has hit the historical record in some places. Taking Beijing as an example, you have mentioned the so-called "blue" during major events and conventions. No major event has been held in Beijing this August but the Beijing blue also appeared. The average monthly concentration of particulate matter, i.e., $PM_{2.5}$ was 38 $\mu g/m^3$ in August this year, a year-on-year decrease of 19.1%. This is the first time that the monthly average concentration falls below 40 $\mu g/m^3$ in recent years, hitting a historical low.

Moreover, the percentage of days of excellent air quality reached 74.2% in Beijing this August, marking a year-on-year increase of 19.4%. The aggregate $PM_{2.5}$ concentration in Beijing from January to August registered at 60 $\mu g/m^3$, a year-on-year decrease of 4.8% and a reduction by 35.5% as compared to 2013.

You have just mentioned the convening of the 19th National Congress of CPC, which is of enormous significance in the decisive stage in building a moderately prosperous society in all respects and the key period in the development of socialism with Chinese characteristics. This is a pivotal event in the political life for all Chinese people.

On July 26th, President Xi Jinping delivered an important speech at the opening ceremony of Workshop on Studying the Key Principles of President Xi Jinping's Speech and Prepare for the 19th National Congress of CPC, in which he expounded on the historical transformation of the undertakings of the Party and the country during the past five years, pointing out that we must unswervingly pressed ahead with ecological civilization and stride forward on the path of building a beautiful China, emphasizing that we must anchor the people's aspiration for a better life and win the war against pollution.

Since the 18th National Congress of CPC, the CPC Central Committee with President Xi Jinping at the core has regarded ecological civilization a key element in the "five-in-one"[2] master plan and "comprehensiveness in four areas"[3] in a coordinated way, which is put at a strategic position in the governance of the country. Plans were developed by the Central Committee for a series of fundamental, far-sighted and groundbreaking undertakings and redoubled efforts were taken by all regions and authorities, which have brought about a historical and holistic transformation in the understanding and practice of ecological environmental protection in China. The five years since the 18th National Congress of CPC are the best period for ecological civilization building and eco-environmental protection in China in terms of in-depth awareness, dynamic actions, effective measures, rapid progress, and heartening results.

With strong determination, we shall align our thinking and actions with the decision and arrangement of the CPC Central Committee and the State Council, secure victory on the main battlefield of eco-environmental protection and conscientiously take the political responsibility of ecological civilization building and eco-environmental protection. We shall make vigorous efforts in shaping a green development model and lifestyle, focusing on tackling the most salient issues arousing public discontent such as air and water pollution and soil contamination in the overall environmental pollution prevention and control with deepened reforms in eco-environmental protection and improvements in the institutional system of environmental protection so as to accomplish the toughest tasks in eco-environmental protection.

Actually, we have already answered your question during the press conference on September 1st. Ten ministries and commissions including the MEP and governments of six provinces and municipalities have initiated the Action for Comprehensive Control of Air Pollution during Autumn and Winter in Beijing–Tianjin–Hebei Region and Its Periphery Areas, addressing the weakest links in air

[2]Coordinated progress in economic, political, cultural, social and ecological areas.

[3]Building a moderately prosperous society in all respects, deepening reform, fully implementing the rule of law, and strengthening party discipline.

pollution control through "1+6" [4]action plan, which is compared to combination blows to bring about continuous improvements in regional atmospheric environment. The action is now progressing as planned and we will keep the public informed about the progress in a timely manner. The MEP shall continue its efforts and break new grounds in preparation for a victorious 19th National Congress of CPC with heartening achievements. Thank you.

Southern Metropolis Daily: Recently, Schaeffler claimed that its raw material supplier was suspended from production due to "environmental reasons," which may result in consequential losses of 300 billion RMB. Some people believe that emphasizing on environmental protection will affect economic development, and some even claim that price hikes of some raw materials are related to environmental protection inspections. What is your viewpoint?

Cui Shuhong: I would like to take this question. We do not agree with the claim that emphasis on environmental protection would hamper economic and social development and cause price hikes in raw materials.

Since the 18th National Congress of CPC, the CPC Central Committee and the State Council have made a series of arrangements for the coordinated development of environment and economy. President Xi Jinping emphasizes that man and nature are in a symbiotic relationship and environmental damage would eventually jeopardize the survival of human beings. We need to put ecological environmental protection in a holistic, long-term, and comprehensive context. President Xi Jinping made the famous statement that "green is gold." Both statements are the most profound, vivid and lively elucidation of the relationship between environmental protection and economic development. We must reach a common understanding and take synergetic actions in the mandate of the CPC Central Committee and the State Council and in the key principle expounded in President Xi Jinping's speech, pressing ahead with environmental protection according to law, boosting the continuous industrial restructuring and promoting the coordinated and sound economic and social development.

In the past, there were some enterprises illegally discharging pollutants with major cuts in or simply zero investment in environmental protection, as what we call "bad money drives out good money," severely disrupting the market competition and hampering the optimization and upgrading of industrial structure. Rigorous law enforcement solves some environmental problems in daily life and improves environmental quality, while giving impetus to economic restructuring and accelerating the renewal of growth drivers.

Regulation and supervision of pollutant-discharging enterprises according to law has become commonplace, to which all enterprises should be adapted and raise their self-consciousness so as to respect and abide by environmental protection laws and regulations.

---

[4] *《Action Plan for Comprehensive Campaign against Air Pollution during Autumn and Winter in Beijing, Tianjin and Hebei and its Periphery Area from 2017 to 2018(hereinafter referred to as "Plan") and its six supporting plan with stepping up inspection, follow-up inspection, targeted inspection, quantified accounting, information disclosure and publicity.*

As for whether environmental protection inspections have affected economic and social development and incurred price hikes of raw materials, I would like to share with you some statistics and two examples, which should be convincing enough. I believe that you are quite familiar with these statistics published by the competent authorities. First, from January to August, the value-added of industrial enterprises above designated size registered at 6.7%, a major increase as compared to the same period last year. 9.74 million urban new jobs were created, a year-on-year increase of 260,000. The registered unemployment rate by the end of the 2nd quarter was at a 3.95%, lowest since 2008. Second, the value-added of industrial enterprises above designated size saw a year-on-year increase of 0.95% in the 15 provinces subject to the central environmental inspection; while the increase was 0.54% in the 16 provinces where such inspections were not yet carried out. Third, from January to August this year, except for cement which underwent a year-on-year decrease of 0.5%, the output of all major industrial products remained stable with moderate growth as the output of crude steel, nonferrous metals, coke, plate glass, and paperboard, respectively, grew by 5.7%, 4.2%, 4.0%, 3.0%, and 1.7%, and the growth rate of crude steel, nonferrous metals, coke, plate glass, and paperboard in the 15 provinces subject to central environmental inspection was above the country's average.

I would like to make four points based on these statistics. First, China's economic and social development is still on the track of steady growth with progress and sound momentum. Second, changes in prices are mainly caused by supply–demand relationships and are determined by the general situation of the current economic development, resource endowment and market demands, etc. Third, price hikes aggravated by exaggerating the environmental policy factor under the pretence of environmental protection is an act of market disruption. Fourth, environmental inspections have not affected the output of major industrial products, nor do they constitute an instant catalyst for price increase.

The significance of environmental protection inspection should be reflected in the mounting pressure of environmental protection to drive out the "bad money," to create a more regulated market environment and to facilitate economic restructuring, thus improving the quality of economic development and safeguarding people's environmental rights and interests.

Here I have two examples of personal experience. Recently I went to Xingtai in Hebei Province and Jining in Shandong Province for the communication and implementation of the principles elaborated by Mr. Li Ganjie, Minister of MEP in Workshop on Action Plan for Air Pollution Prevention and Control during 2017–2018 Winter. I found that all 1,387 sheet material manufacturers without pollution control facilities and causing severe environmental pollution were demolished altogether. You may visit the demolition sites and you will be impressed. Besides eradicating the pollution sources, the factory buildings were pulled down as well. At the same time, a new industrial zone reaching internationally advanced level is planned for construction to encourage high-quality international enterprises to settle in while guiding the banned enterprises to move back in after technical improvement and quality upgrade to rebuild modernized and environmentally friendly sheet material enterprises.

In the first half of this year, the increase of $PM_{2.5}$ concentration in Xingtai was lower than the average of Hebei Province and the improvement rate ranked the third in the province. More importantly, from January to August this year, the growth rate of value-added of industrial enterprises above designated size in Xingtai ranked the top of the province, and the increment and growth rate of total fiscal revenue and public budget revenue both ranked the second. The example of Xingtai shows that a win-win situation of economic growth and environmental protection can be achieved provided that appropriate measures are taken, making environmental protection a powerhouse for accelerated transformation of growth drivers. This is a positive example which gives us confidence.

Another example is Jining in Shandong Province, which is a heavy industrial city with 10.25 million kw of installed capacity of thermal power and 80 million tons of coal mining, of which nearly 50 million is locally consumed, together with steel, coking and coal chemical plants. Jining has implemented decisive measures on the poorly managed enterprises that are in unauthorized locations, lack proper certificates and fail to meet emission standards while practicing precision management for pollutant-discharging enterprises, achieving encouraging results of air pollution control. From January to August this year, the concentration of $PM_{2.5}$ registered at 55 $\mu g/m^3$, a year-on-year improvement of 22.5%, and both indicators ranked among the top in the "2+26" cities with evident decrease and improvement as compared to its "2+26" peers.

From an economic perspective from January to June, the GDP growth in Jining hit 7%; the main business revenue growth of industries above the designated scale was 0.66% higher than the provincial average; the profit growth exceeded the average level by 28.9% while tax revenue was 23.7% higher than average. According to the governmental officials of Jining, there was no silver bullet at all; it was rigorous management according to law that produced the good results and the win-win situation of environmental protection and economic development.

There are many other examples in China similar to Jining and Xingtai. Despite the enormous tasks remaining on the agenda of environmental protection, these two examples have given me confidence that what is accomplished in these cities can be done elsewhere too. If time permits, I would encourage you to visit these cities, as seeing is believing. I was greatly inspired and delighted during my visit. Thank you.

Guangming Daily: I have two questions. First, why did you choose Lianyungang, Chengde, Ordos, and Jinan in compiling the technical guide for the line, limit, and list? What is so typical about them? Second, I have read about the Checklist and Inspection List for "Green Shield 2017" Special Supervision and Inspection Action of National Natural Reserves. Could you elaborate on that? How is the Green Shield Special Action progressing now?

Cui Shuhong: I have briefly introduced the line, limit and list, which is on the top of the agenda of environmental management and EIA reform. The reason we have chosen Lianyungang, Ordos, Chengde, and Jinan for pilot project is closely related to the background of the line, limit and list. We all know that Lianyungang is a prioritized city and area in the coastal region of Jiangsu Province, where the port development, coastline development, and chemical plants would affect the regional environment. Therefore it is worthwhile to seek solutions to optimize coastline development and determine the maximum limit of pollutant emissions through the line, limit, and list. In the pilot project in Lianyungang, the land resources were delineated into over 200 control areas by setting the line, limit, and list, with specific requirements for environmental quality and resource development and utilization of each control area. This is a significant move in the healthy economic and social development of Lianyungang down the road.

Ordos is an energy-oriented city with high intensity of resource exploitation and utilization. Industries associated with resource exploitation, including coal chemical industry, show strong momentum of development. This city is under tremendous pressure of resource exploitation and environmental protection. The line, limit, and list have defined the maximum threshold for the resource exploitation intensity and total emissions from the coal chemical industry, which means a lot for this region.

Considering the strong ecological protection function of Chengde, setting the red line of ecological protection, the bottom line of environmental quality and the maximum limit of resource utilization means a lot for the sustained cleanliness of this region in long term.

Jinan is a provincial capital with high intensity of economic and social development; thus achieving equilibrium in economic and social development and environmental protection in such a provincial capital would set a good example for other provincial capitals. Our choice of these typical cities for pilot work proves to be quite effective up to now.

"Green Shield 2017" is a major action taken by environmental protection authorities in response to the requirement by the CPC Central Committee and the State Council to learn lessons from the ecological problems arising in Qilian Mountain Natural Reserves and to raise awareness of the political implications of eco-environmental issues. To initiate this action, the MEP has convened a nationwide video conference for planning and arrangements, requiring all provinces (autonomous regions and municipalities) to clean up, screen and take necessary remedial measures for development behaviors in natural reserves in their respective jurisdiction by the end of this year. Based on what I know, the MEP is planning for another meeting for remobilization, rearrangements, and redeployment.

Liu Youbin: That's all for today's press conference. Thank you.

Regular Press Conference (October)

October 31, 2017

On the morning of October 31, 2017, the MEP held the October press conference, introducing the progress of some key work of the MEP. Fang Li, Director-General of the Beijing Municipal Environmental Protection Bureau, Wen Wurui, Director-General of the Tianjin Municipal Environmental Protection Bureau, and Gao Jianmin, Director-General of the Hebei Provincial Environmental Protection Administration, respectively, talked about the progress of the key action of comprehensive treatment against air pollution during the fall and winter seasons in the Beijing–Tianjin–Hebei region and answered questions from the media. This is the first time that the major responsible persons from local environmental departments are invited to such press conferences.

Ministry of Environmental Protection of the People's Republic of China,
2017 Press Conference Records of Ministry of Environmental Protection, the People's Republic of China, https://doi.org/10.1007/978-981-13-7330-5_11

The MEP October Press Conference (1)

The MEP October Press Conference (2)

Liu Youbin (Moderator): Dear friends from the press, good morning! Welcome everyone to the MEP October Press Conference.

General Secretary Xi Jinping pointed out in the report of the *19th National Congress of the Communist Party of China* that "get everyone involved in improving the environment and address environmental issues at the root. Continue the campaign to prevent and control air pollution to win the battle for blue sky." General Secretary Xi Jinping attached great importance to the effort of air pollution prevention and control, which provided us with the determination and confidence in the prevention and control of air pollution.

Since this September, the MEP collaborated with 10 departments and 6 provinces or cities to launch the key action of comprehensive treatment against air pollution during the fall and winter seasons around the Beijing–Tianjin–Hebei region. It is an essential step for making our skies blue again. In today's press conference, we have Mr. Fang Li, Director-General of the Beijing Municipal Environmental Protection Bureau, Wen Wurui, Director-General of the Tianjin Municipal Environmental Protection Bureau, and Gao Jianmin, Director-General of the Hebei Provincial Environmental Protection Administration to talk about the progress of the key action in the Beijing–Tianjin–Hebei region and answer your questions.

Next, I would like to briefly introduce the progress of some key work in the MEP recently.

1. The MEP quickly put the guiding principles of the 19th National Congress of the CPC into action

From October 18 to 24, the historic 19th National Congress of the CPC was held in Beijing successfully. After the closing of the conference, the Party Secretary and Minister of the MEP, Li Ganjie, promptly presided over the leading party group meeting and the (enlarged) leading party group meeting to convey the grandness of the 19th National Congress of the CPC and the guiding principles of the report of the 19th National Congress of the CPC, the Work Report of the 18th Central Commission for Discipline Inspection, and the Party Constitution Amendment, and instructed all organs and institutions within and affiliated to the MEP to actively study the guiding principles of the 19th National Congress of the CPC.

From October 29 to 30, the MEP convened a seminar themed "Study and put the guiding principles of the 19th National Congress of the CPC into action to win the battle for a good eco-environment," to convey the guiding principles to the national environmental protection system, trigger thoughts and exchanges, as well as study and deploy implementation work. Responsible persons of the environmental protection departments of various provinces (regions and municipalities), cities under separate state planning, and Xinjiang Production and Construction Corp, cadres at the Division-Head level or above from the departments of the MEP, main responsible persons of the dispatched offices and directly affiliated agencies of MEP attended the meeting.

Minister Li Ganjie pointed out that studying and implementing the guiding principles of the 19th National Congress of the CPC is the primary political task of the environmental protection system for now and in the near future. It is necessary to understand, master, and fully implement the guiding principles and closely connect it to ecological environmental protection efforts, to ensure that the guiding principles of the 19th National Congress of the CPC can truly embed in the environmental protection system and generate fruitful results. We need to focus on the subject of fighting for a good eco-environment, enhancing ecological civilization, and building a beautiful China, while implementing the important strategy on ecological civilization of General Secretary Xi Jinping, as well as concentrate on the three objectives of promoting a green development and lifestyle, upgrading the ecological environmental quality, and modernizing the system and capability of the national eco-environmental control sector. We should stay focused on the following 6 key tasks:

1. Establish and strictly abide by 3 redlines. Strictly abide by the baseline to ensure ecological functions, the bottom line of environmental quality safety, and the cap line of using natural resources.
2. Promote green development. Comprehensively optimize industrial distribution, accelerate industrial structure adjustment, and minimize the resource consumption of productions, pollution emission density, and total volume.
3. Effectively solve distinctive environmental issues. Resolutely make our skies blue again. Launch the clean water action. Firmly carry out the land purification campaign. Comprehensively promote the environment in rural area and effectively prevent and control environmental risks.
4. Accelerate ecological protection and restoration. Implement the key projects of important ecological system protection and restoration, establish ecological corridors and biodiversity protection networks, and reinforce the national ecological safety barrier.
5. Carry out green initiatives in public society. Actively leverage the role of the government as the guidance, improve the self-disciplinary mechanism of enterprises, and encourage public participation, in order to build up an environmental protection system with the government as the lead, enterprises as the main body, and civil societies and the public as participants.
6. Deepen the reform of the eco-environmental protection mechanism. Reform the eco-environmental supervision system. Improve the environmental protection inspection mechanism, promote the inspection system at the central and provincial level, and enhance the assessment and accountability of eco-environmental protection. Improve the guiding and incentive mechanism of eco-environmental protection.

Minister Li Ganjie said that the 19th National Congress of the CPC made new arrangements for the in-depth promotion of the new great projects of the Party building. The environmental protection system nationwide should firmly implement the requirements for a full and strict governance over the party, raise the awareness, and learn it in-depth. It's also required to seriously implement the general requirement of Party building, and ensure strict Party self-governance.

Minister Li Ganjie emphasized that we need to improve our effort and speed of implementing the requirements of the upcoming National Environmental Protection Conference with a positive mindset and strong capability, aiming at high standards, requirements, and quality, to take the upcoming National Environmental Protection Conference as a channel to convey the guiding principles of the 19th National Congress of the CPC and an opportunity to mobilize the joint effort of building a beautiful China.

2. The key action of the comprehensive treatment against air pollution during the Fall and Winter seasons achieved positive results.

Since the launch of the action on comprehensive air pollution prevention and control in the autumn and winter of Beijing–Tianjin–Hebei region and its periphery for 2017 to 2018, "2+26" cities mainly focused on the key industries of coal, manufacturing, and automobile to take actions in advance and accelerated the implementation of various treatments. Everything is going quite well. From October 1 to 27, the average concentration of $PM_{2.5}$ in the "2+26" cities in the Beijing–Tianjin–Hebei region and surrounding areas was 62 $\mu g/m^3$, down by 4.6% than a year ago. The air quality in this region steadily improved.

First, we promoted the clean heating project in winter. By the end of September 2017, we used clean fuels to replace small-scale "scattered" coal burning in 2 million households, replaced nearly 4 million tons of coal.

Second, we adopted the measure of simultaneous investigation, reconciliation, and rectification while enhancing the strength of treating the companies that are "in unauthorized locations, lacked the proper certificates and failed to meet emissions standards." The organization deeply implemented the comprehensive control of the companies and clusters that are "in unauthorized locations, lacked the proper certificates and failed to meet emissions standards" and adopted network-based inspection. From September 1 to October 22, the supervision group investigated 69,671 points on-site, which had implemented comprehensive rectification to the companies with "location, certificate, and emission" issues, phased out or remodeled coal-fired boilers, and treated VOCs. It is found out that 5,978 points still had environmental issues. The group instructed local governments to conduct rectification.

Third, we continuously enhanced the comprehensive rectification of industrial companies, promoted rectification and management on unorganized emission and VOCs treatment, and accelerated the management of emission permits of key industries. By the end of September 2017, the "2+26" cities phased out more than 50,000 coal-fired boilers, remodeled 803 boilers that equal to 50,000 t/h, and finished the comprehensive treatment of VOCs of 3,866 companies.

Four, we strengthened the supervision of vehicles and industrial equipment that exceeded emission standards, seriously investigated excessive emission of heavy-duty trucks, and improved the supervision and management of the oil for automobiles.

Five, we established the off-peak production plan for the industries of iron and steel, coking, foundry, building materials, nonferrous metal, chemicals, etc., specified the list of off-peak production companies, and carried out off-peak transportation based on the requirement of production.

For the next step, the MEP will urge local departments to complete tasks defined by the stepped-up action on time, speed up air pollution control, and ensure that the average concentration of $PM_{2.5}$ in the "2+26" cities between October 2017 and March 2018 can decrease by 15%, while the number of heavy pollution days can drop by over 15% than the same period of last year.

3. Solidly promoted "Green Shield 2017", the special action of supervision and inspection of the national nature reserves.

This June, the General Offices of the CPCCC and the State Council printed and distributed the *Circular on the Inspection of and Solutions to the Ecological and Environmental Problems in Gansu Qilian Mountain National Nature Reserve and the Lessons Learned*. In order to take the guiding principles of this document into action, the MEP collaborated with 6 departments and kicked off the "Green Shield 2017" special action of supervision and inspection of the national nature reserves to strictly crack down on illegal activities and violations related to nature reserves.

Firstly, we carried out comprehensive supervision and inspection. 7 departments including the MEP held video conference for the special action, printed and distributed the work plan of this action, and established cross-department coordination groups, to conduct all-around supervision and inspection of various illegal acts and violations in national nature reserves. Provinces established multi-sectoral joint working mechanisms and groups, carried out and distributed the work plan for the special action and deployed work accordingly. Nature reserves at city and county level adopted both self-inspection and supervision from provincial working groups to clarify responsibilities on different layers.

Secondly, we solidly implemented verification. According to incomplete statistics, 20 provinces have completed the on-site inspection of provincial working groups according to the requirement of the special action and conducted verification referencing the problem list from remote sensing monitoring. At the same time, issues related to nature reserves discovered through central environmental protection inspections, nature reserves supervision and inspection, and the ones covered by admonitory talks and direct supervision in recent years, as well as the problems newly found in this inspection, are uniformly included in the management book of illegal activities and violations for further investigation and treatment. Up to now, we have found more than 30 local regulations that are not aligned with host laws.

Thirdly, we seriously investigated violations and penalized responsible parties. According to incomplete statistics, up to now, local departments have investigated and dealt with more than 7000 issues related to nature reserves and carried out rectification suggestions and requirements. Some issues have been solved and ecological restoration measures have been implemented. In terms of accountability, local departments have held over 300 responsible persons accountable. In particular, the Party Committee Government of the Xinjiang Uygur Autonomous Region seriously investigated and penalized a number of leaders on the violation of adjusting the scope of the Karamali Mountain Nature Reserves for 6 times, fully exerted the effect of deterrent, warning, and education.

Fourthly, we enhanced public supervision. Local departments fully leveraged the internet, TV, newspaper, and media like MicroBlog and WeChat to actively cover related reports on the Green Shield Action and create a positive atmosphere to attract public participation in protecting nature reserves.

From October 25, the MEP has been conducting a follow-up inspection of the special action together with other 6 departments to review the discovered problems and track the progress. We gathered about 200 people from relevant departments and divided them into 10 teams to conduct a one-month follow-up inspection in 31 provinces. We are going to have admonitory talks or special direct supervision to the local government and management department of cities and counties that did not fully investigated or rectified discovered issues, hold them accountable to ensure the effectiveness of the special action.

Next, please invite Director-General Fang Li to share with us some information.

Fang Li: Dear friends from the press, good morning! I am very pleased to meet you all. First of all, I would like to thank you for your concern and support to the environmental protection in Beijing. Next, I will briefly introduce the progress of air pollution prevention and control in Beijing.

From the start of this year, under the lead of the Party Central Committee and the State Council, and the strong guidance of the MEP, Beijing unswervingly followed the guiding principle of socialism with Chinese characteristics in the new era raised by General Secretary Xi Jinping, resolutely put the people-centered development strategy in practice, consciously established and implemented the concept of "Lucid waters and lush mountains are invaluable assets," to implement air pollution control measures beyond the usual state based on the principles of the speeches given by General Secretary Xi Jinping during his 2 visits in Beijing. Particularly, we implemented the principles of the 10th Meeting of the Coordination Group of the Air Pollution Prevention and Control in Beijing–Tianjin–Hebei region and its periphery, carried out the plan of the key action in the fall and winter seasons, and held deployment meetings. Party Secretary of the Municipal Party Committee, Cai Qi, and the Acting Mayor, Chen Jining, clarified the disciplines and requirements to the Party Committee governments and related departments at all levels and urged them to pay close attention to the implementation.

Fang Li, Director-General of the Beijing Municipal Environmental Protection Bureau

The key action plan of Beijing has 30 measures and 3 characteristics. One: face the problem directly; two: strictness is the priority; three: implementation is the bottom line. Specifically, it means:

Enhance pollution control and emission reduction measures. We have finished the remodeling of switching more than 10,000 t/h of coal-fired boilers to use clean energy, the low-nitrogen treatment of 19,000 t/h gas-fired boilers, significantly exceeded the annual plan ahead of schedule. We have also overfulfilled the "coal-to-clean energy" task in 700 villages. Within this year, 6 districts in the municipality and the southern plain area will basically be "coal-free" and high-quality energy will account for about 90% of the total energy consumption. 375,000 vehicles were scrapped, 624 companies were phased out from the general manufacturing industry or closed as polluters, both exceeded the annual target. All of the taxis with Three-way Catalytic Converters for more than 2 years had their TWCs changed, companies with the "location, certificate, and emission" issues on the record were all closed.

Promote the strictness of political standards. We revised and released the production process and equipment phase-out catalog of the pollution-intensive industries, which now includes 172 items. According to the requirement of the MEP, we revised the emergency response plan of heavy air pollution again and implemented it strictly with the highest standards. We inspected 22 key companies in the thermal power industry and issued emission permits. "Jing VI" standard gasoline and diesel are supplied in Beijing. In October, the quality of ordinary diesel and automobile-used diesel are aligned. Beijing collaborated with Tianjin and Hebei to establish the first unified standard of the building coatings' VOCs for a better management and control throughout the entire process.

Enhance the pressure during law enforcement for a better deterrent effect. From January to October, we focused on the issues of companies with the "location, certificate, and emission" problems and heavy-duty diesel vehicles, implemented full-time precise law enforcement. The penalty for fixed pollution source exceeded 160 million RMB, with a year-on-year growth rate of 77%. The *Environmental Protection Law* was implemented to investigate and deal with 724 major environmental violations. "Environmental Protection Police" was set up to strengthen the connection between administrative law enforcement and criminal justice, jointly investigated and handled 161 cases. We adopted the new model of "penalty defined by the Public Security department, and evidence collected by the Environmental Protection department" and penalized more than 16,000 heavy-duty diesel vehicles. All major illegal cases were publicly exposed to ensure the deterrent effect of each case.

Ensure the implementation at the community level. Promote the execution capability of organizations at the street, village, and township level for "the last mile" through methods of ranking, admonitory talks, and supervision. We finished the city-level environmental protection inspection to 7 districts, had admonitory talks with 800 people, and held them accountable. Inspection will be expanded to all 16 districts within this year. Based on the high-density monitoring network of $PM_{2.5}$, the air quality ranking was made on a monthly basis at the street, village, and township level, and conducted admonitory talks with the ones with lower ranks or did not fully implemented all treatment measures. The consciousness and willingness of pollution treatment in streets, villages, and towns have been significantly improved.

Strengthened the effort to attract more public participation. The framework of public participation in the "joint treatment" of air pollution is emerging through voicing ourselves, responding to public concerns and doubts, promoting popular science dissemination, and launched the campaign of reporting illegal acts with a prize.

For the next step, Beijing will thoroughly study and implement the guiding principles of the 19th National Congress of the CPC, remain true to our original aspiration and keep our mission firmly in mind, solidly establish the socialist ecological civilization concept, take the citizens' demand for good air quality as a political responsibility and an honorable task, set a high standard and achieve fruitful results in the key action against air pollution in the fall and winter seasons and make our skies blue again, make our contribution from the environmental protection sector to build up a world-class harmonious and livable city.

Last but not least, I sincerely hope our friends from the press can always care about, understand, and support our work. Let's make our skies blue again together. Thank you very much!

Liu Youbin: Next, let's welcome Director-General Wen Wurui to share with us some information.

Wen Wurui: Dear friends from the media, good morning! I am very pleased to have the opportunity to come to Beijing and meet with you all. General Secretary Xi Jinping stated that we should continue the campaign to prevent and control air pollution to make the sky blue again in the report of the 19th National Congress of the CPC. As one from the environmental protection sector, I was truly encouraged and very excited. I would like to take this opportunity to briefly introduce the stepped-up action of air pollution prevention and control in the fall and winter seasons in Tianjin.

Wen Wurui, Director-General of the Tianjin Municipal Environmental Protection Bureau

The municipal Party Committee and the government of Tianjin firmly implemented the decisions made by the Party Central Committee and the State Council to fully push forward the Clean Air Campaign. We controlled the pollution growth through structural adjustment, reduced pollution volume through project measures, prevented regional transfer through joint prevention and control, cut the peak volume through emergency response, addressed both the symptoms and root causes through innovative regulations, and clarified pollution treatment responsibilities through law enforcement and accountability. The air quality is getting better year by year. The concentration of $PM_{2.5}$ in 2016 dropped to 69 μg/m^3, decreased by 28.1% than in 2013.

Aiming at the significant issue of air pollution in the fall and winter seasons, the municipal Party Committee and the government of Tianjin took the task of ameliorating the air quality in the fall and winter seasons as an essential step to truly enhance the sense of fulfillment and happiness for the public, implemented comprehensive policies sparing no effort despite all the difficulties and obstacles, and made some impressive achievements.

First, solidly implemented "zero" small-scale "scattered" coal burning. According to the requirement of "choose electricity or natural gas for heating based on the reality of different households" raised by General Secretary Xi Jinping, we inspected a total of 1.21 million bulk coal users. The "zero" bulk coal campaign can be finished in 2 years instead of 3, and the related work for 610,000 households can be finished this year.

Second, resolutely rectified coal-fired boilers. We found more than 11,000 coal-fired boilers in Tianjin this year and shut most of them down except 184 boilers that reached special emission limit or met extremely low emission requirements.

Third, resolutely managed companies that are "in unauthorized locations, lacked the proper certificates and failed to meet emissions standards." We inspected a total of 19,000 companies with "location, certificate, and emission" issues and implemented different regulations according to different types, halted their production before further rectification. We shut down more than 9,000 companies that had severe pollution problems with no capability for treatment, met the requirement of "cutting industrial water and electricity, and clean the material, product, and production equipment."

Four, resolutely rectified key industries. Deeply involved in the rectification of 452 key companies with VOCs issues, treated unorganized emission of 181 companies in the steel and foundry industries.

Five, firmly implemented off-peak production and transportation. 396 key companies in 10 industries including foundry and steel adopted off-peak production; companies in the coking industry prolonged the coking time to 48 h; 27 companies with high transportation demand realized off-peak transportation.

Six, resolutely treated "pollution related to vehicle, gasoline, public transportation, and port." This year, a total of 40,000 vehicles were culled out, heavy-duty diesel vehicles below the "National Stage III Standard" all had DPF installed. We prohibited the use of mid-size and heavy-duty diesel trucks within (on) the Outer

Ring. We are supplying gasoline and diesel qualified as "National Stage VI Standard" all over Tianjin. The Port of Tianjin realized connected shipment between ocean and railway for cargos and bulk coal, which reduced coal shipment via vehicles by 60 million tons.

Seven, firmly prevented and controlled nonpoint source pollution, including dust. From October 1, construction work on earth and stones has been stopped within the established area in Tianjin. We are strictly monitoring the specially authorized construction projects related to public livelihood. Through adopting technologies of "internet + monitoring cameras," we can monitor crop straw burning in all agricultural areas, send out alarms automatically, and track down the responsible person according to the location.

Eight, focused on the response in heavy pollution days. Beijing, Tianjin, and Hebei adopted joint prevention and control and implemented "different regulations for different plants" to industrial companies to cut the peak pollution volume.

Nine, strictly treated pollution. Cadres at the Director General level from the municipal Party Committee and the government took the lead, supervisory inspection groups are dispatched to be based in districts for supervision. Quantified accountability regulations were printed and distributed to penetrate the pressure to the grass root organization. This year, we have held 1,414 people accountable for environmental protection issues, in which 27 people used to be at Director General level, 636 at Division Director level. Recently, the environmental protection department and the public security department jointly implemented law enforcement, which distinctively enhanced the deterrent effect. Penalty related to environmental protection reached 249 million RMB, about 3 times of that in the same period of the last year.

Ten, resolutely strengthened dispatching and guidance. Keep clear records of the progress and result of key tasks, assign tasks to districts, towns, villages, and households. District governments are dispatching staff for the key action against air pollution on a daily basis, city government on a weekly basis, striving for a success of the key action against air pollution in this fall and winter through the hard work every day, every month, and every season.

Tianjin municipality will fully take the guiding principle of the 19th National Congress of the CPC into action, resolutely make our skies blue again, contribute to the effort of enhancing the air quality in the Beijing–Tianjin–Hebei region. Thank you.

Liu Youbin: Next, let's welcome Director-General Gao Jianmin to share with us some information.

Gao Jianmin: Good morning! I am Gao Jianmin, the Party Secretary and Director-General of the Hebei Provincial Environmental Protection Administration. Thanks for your long-term concern and support to the environmental protection effort of Hebei Province.

I listened to General Secretary Xi Jinping's exciting and encouraging report at the site as a member of the 19th National Congress of the CPC. Particularly, it was defined in the report that the harmonious relations and coexistence of human and nature are one of the essential fundamental strategies of upholding and developing the socialism with Chinese characteristics in the new era. In addition, it strategically deployed the key tasks of accelerating the institutional reform for ecological civilization and building up a beautiful China, which set the direction and target for us to promote the ecological conservation and environmental protection.

Recent years, the provincial Party Committee and the government of Hebei firmly implemented the guiding principle and requirements of General Secretary Xi Jinping about ecological civilization and air pollution prevention and control. With our tenacious courage and the determination of fighting an uphill battle, we firmly pushed forward the comprehensive air pollution control, which led to significant improvement of the air quality, met the target of reducing $PM_{2.5}$ by 25% set by *The Action Plan on Air* 2 years in advance. I want to take this great opportunity to give you a brief introduction on the comprehensive treatment and the stepped-up action in the fall and winter seasons against air pollution made by Hebei Province.

Organization development: focusing on treating haze scientifically, collaboratively, and strictly while keeping the top-level organization development in mind, mobilizing all forces and resources from the Party and the entire society to succeed in the stepped-up action of a comprehensive treatment against air pollution. *The Opinions on Firmly Pushing Forward the Comprehensive Air Pollution Control in Hebei* and 18 auxiliary implementation plans were established through the study and plan personally conducted by the leaders of the Provincial Party Committee and the government, direct guidance and strong push from provincial leaders in charge of different sectors, and the participation and cooperation of 29 departments directly under the leadership of the provincial government. A new pattern for promoting the comprehensive air pollution control was established with the participation of departments at all levels in the provincial government and people from all walks of life in the society. At the same time, based on the characteristics of the air pollution in fall and winter, we established the *Action Plan on Comprehensive Air Pollution Control in the Autumn and Winter of Hebei Provinces for 2017 to 2018*, as well as 5 supporting plans for targeted inspection, law enforcement supervision, quantified accountability, information disclosure, publicity, covering 6 aspects, 13 items and 40 detailed strengthen measures, established stricter mechanisms for supervision and inspection, accountability, information disclosure, and the quantified performance assessment of the departments under direct leadership by the provincial government and city and county governments. With the firm resolution, strict discipline, strong mindset, and serious measures, we declared the war against haze, sparing no effort to make our skied blue again.

Gao Jianmin, Director-General of the Hebei Provincial Environmental Protection Administration

Treatment measures: focusing on treating the "symptom" in summer and the "root cause" in winter, prioritizing the structural issues of heavily relying on heavy industries, fossil fuels, and road transportation. Up to now, 33,600 small-sized coal-fired boilers (equals to 50,500 t/h) were culled out, reduced iron production capacity by 20.22 million tons, steel production capacity by 22.61 million tons, concrete by 2.015 million tons, plate glass by 5 million weight cases, coking by 8.08 million tons. Reduced the coal consumption by 38.45 million tons, with the major pollutant emission of SO_2, NO_x, and smoke dust dropping by 128,400, 152,600, and 175,800 tons. We completed the in-depth rectification of 1,227 companies with VOCs issues, reduced its emission by 183,000 tons; scrapped 219,000 outdated vehicles, which reduced the emission of NO_x by 17,000 tons. Heating by clean energy: completed "replacing coals with gas" and "replacing coals with electricity" in 1.812 million households, shut down 2,796 bulk coal selling sites and effectively controlled the pollution caused by inferior coal firing; rectified 108,500 companies with problems including missing pollution treatment equipment, having substandard equipment, or "location, certificate, and emission" issues exceeding emission standards, effectively managed and controlled companies that were not included in statistics, conducted extensive management of emission control or direct discharge, and significantly reduced the emission of various pollutants. On the other hand, remodeled and relocated 31 heavy pollution companies, phased out 8.23 million tons of coal production capacity, finished the rectification of 4,685 construction projects and 305 outdoor mines, restored green landscape at

140 sites and 5.727 million mu of forestation, coordinately pushed forward both pollution control and ecological restoration. From October 1 to 24, $PM_{2.5}$ dropped by 25.7% on a year-on-year basis which represents significant results of the stepped-up action against air pollution in fall and winter.

Implementation: pushing forward the inspection of governments and companies at the same time, continuously implementing supervision on targeted projects and law enforcement, focusing on solving the issue of the "last mile" of haze treatment. We enhanced the strictness of accountability, seriously handled and penalized the responsible persons who did not seriously implement the rectification requirements instructed by the MEP and the provincial government, practiced fraud on the implementation, violated regulations for multiple times, or refused to conduct rectification. By October 25, there are 10,297 cases filed due to the incompetence of implementing measures of haze treatment, 1,563 persons received the Party's administrative punishment, 78 persons were removed from their original posts, deposed, or demoted. 697 cadres were involved [in which 77 were at the County level (Division level)], and 42 leading Party groups were held accountable. During this fall and winter, we implemented 2 rounds of targeted inspection and law enforcement supervision. A total of more than 51,200 person times were dispatched for the inspection of 1,368 governments at various levels, inspected 46,000 companies and discovered 9,262 environmental issues of various types. We filed 1,368 cases and 79 persons were transferred to the public security department for accountability. Penalized 81 cadres with the Party's administrative punishment and effectively clarified the responsibilities of different departments at various levels to showcase the determination of strictly cracking down on air pollution violations.

Mechanism support: promoting the management and control system, strengthening the standards to reversely force companies to enhance their environmental protection efforts, taking a range of measures to ensure the air quality. We comprehensively implemented the layered responsibility distribution mechanism, which cities shoulder the responsibility of supervising counties, counties supervise towns, while towns supervise villages. We also clarified the responsible persons according to the supervision grid, to create a seamless responsibility system for the environmental protection supervision. We enhanced the management through checklists, conducted all-around inspections, and established checklists of pollution sources, which supported the accurate instruction on off-peak production and emergency response in heavy pollution days. We promoted emission standards and emphasized on the implementation of the limit of ultra-low emission and special emission, drove company transformation and upgrading with both support and pressure, to realize the leap of green development of the economy in Hebei through "Phoenix Nirvana".

Next, according to the new requirements for the ecological civilization and environmental protection stated during the 19th National Congress of the CPC, with continuously promoting air quality as the core, upholding the principle of public participation and prevention and control from the source, we will strengthen the environmental supervision and inspection, implement all related measures, and mobilize the strength and resources of the entire Party, the public, and the society to participate in haze control, resolutely make our skies blue again while providing eco-products with higher quality, bring the sense of fulfillment and happiness of a beautiful eco-environment to all! Thank you.

Liu Youbin: You may ask questions.

China Daily: "Building up a Community of Shared Future for Mankind" was stated in the report of the 19th National Congress of the CPC, and "The Belt and Road Initiative" was involved in the Party Constitution. What have the MEP done for a green BRI?

Liu Youbin: Thanks, I will briefly answer this question. The "Belt and Road Initiative" is an essential initiative for international cooperation raised by the Chinese government. General Secretary Xi Jinping clearly stated during "The Belt and Road" Forum for International Cooperation in May 2017 that "We should pursue the new vision of green development and a way of life and work that is green, low-carbon, circular and sustainable. Efforts should be made to strengthen cooperation in eco-environmental protection and build a sound ecosystem so as to realize the goals set by the 2030 Agenda for Sustainable Development." He also mentioned that China will establish an Environmental Big Data Service Platform and initiate an International Coalition for Green Development on the Belt and Road.

In the report of the 19th National Congress of the CPC, General Secretary Xi emphasized that we need to establish a community with a shared future for mankind, and build an open, inclusive, clean and beautiful world that enjoys lasting peace, universal security, and common prosperity. We also need to uphold the environmentally friendly principle and protect the earth, the only home of human beings. The MEP will seriously implement the important principles, concepts, and requirements of the General Secretary Xi Jinping to push forward the green "Belt and Road Initiative" with strong efforts.

To this end, the MEP mainly focused on 3 tasks:

First, enhanced the organizational structure from the top level. This May, the MEP, the Ministry of Foreign Affairs, the NDRC, and the Ministry of Commerce jointly published the *Guiding Opinions of the Green Development on the Belt and Road*. The MEP also published the *Plan for Cooperation in Eco-Environmental Protection for the "Belt and Road Initiative"* in order to fully implement the *Guiding Opinions*. The Guiding Opinions clearly defined the general direction of the green development on the Belt and Road as a programmatic document. It solidly emphasized the concept of innovative, harmonious, green, open, and shared development, fully promoted the green progress of policy exchanges, infrastructure connection, trade facilitation, capital investment, and people communication. The "Plan" is a practical instruction for the implementation of the "Guiding Opinions", which involves 25 key programs for building up a diversified cooperation landscape of eco-environmental protection.

Second, established a cooperation and communication platform of eco-environmental protection. We launched the "Green Envoys Programme", which triggered policy communications and the sharing of environmental protection practices. More than 700 people have participated in this plan. The "Belt and Road" environmental protection technology communication and transfer center was established in Shenzhen. Together with the UNEP, we launched and organized an International Coalition for Green Development on Belt and Road to help the countries involved in the BRI implement the 2030 sustainable development agenda. Currently, the two parties are drafting the framework and fundamental documents for the coalition.

Third, actively guided Chinese companies to fulfill their environmental protection responsibilities. Pushed major companies to publish and implement the initiative of promoting the green development on Belt and Road and act in compliance with the environmental laws and regulations of their own countries consciously. Many projects involved in the "Belt and Road Initiative" are following the environmental protection standards that are much stricter than the local requirements. Take the coal-fired power plant in Sahiwal, Pakistan contracted by a China enterprise for an example, the SO_2 and NO_x emissions are far lower than the local standard. The China-contracted thermal power project in Cuddalore, India received the environmental protection award from an Indian foundation on promoting full-scale power generation.

Next, the MEP will continuously expedite the construction of the "Belt and Road" Environmental Big Data Service Platform. We will provide eco-environment information, green supply chain cooperation opportunities, green financial sharing services through leveraging the technologies of big data, cloud computing, satellite remote sensing, etc., to promote the environmental protection coordination among countries and their comprehensive response capabilities.

We have the "heads" from the local environmental protection departments of Beijing, Tianjin, and Hebei with us today, so let's stay more focused on the issues specifically related to these 3 regions. Next question.

CCTV: We know that vehicle emission, especially the heavy-duty diesel vehicles, is the major pollution source of $PM_{2.5}$ in Beijing. I would like to ask that what are the major measures to treat the emission of heavy-duty diesel vehicles in Beijing?

Fang Li: Thanks for your question. As we know, there are several major measures for treating the emission of heavy-duty vehicles. I will briefly talk about them.

There are mainly two types of heavy-duty diesel vehicles in operation in Beijing, one is the heavy-duty trucks that are necessary for the logistics in the city; and the other one is trucks in transit. We can see from the map that a lot of roads are connecting Beijing to the capitals of provinces, which means Beijing has the highest transportation network density. There is a great amount of cargo transportation in transit from the South to the North, and from the East to the West. Heavy-duty vehicles are closely connected to the logistics of production and our livelihood.

So the priority of the emission treatment of heavy-duty vehicles is to ensure that the diesel is getting continuously cleaner. This year, Beijing took the lead and supplied the vehicle diesel of National Stage VI Standard. Frankly speaking, the current standard of vehicle diesel is still lower than the strictest standard in the world. For the next step, we are going to establish higher and stricter standards and make the high-standard diesel available to the vehicles as soon as possible.

Secondly, we will reduce the emission density of individual vehicles. The MEP has announced the National Stage VI Standard and we are expecting that Original Equipment Manufacturers can provide vehicles that meet the National Stage VI Standard as soon as possible. There is going to be a process from the establishment of the standard to the production of new vehicle models, but we hope that the cleaner National Standard VI vehicles can be available sooner. We believe that the new vehicles will be available in the Beijing–Tianjin–Hebei area first.

Thirdly, we need to adjust vehicle structure. However, it is certain that the adjustment cannot be done overnight, it needs time. The safe operation of the city is the priority during the adjustment. Safety is the baseline for anything. With the technology development, we hope that more and more new energy EVs can replace diesel vehicles for logistics use with the improvement of EV capacity to reduce pollution.

Fourthly, we will strengthen the punishment. The environmental protection department now requires all employees to be onboard all time and mainly focus on the heavy-duty diesel vehicles that exceeded the emission standard to strictly crack down on such misconducts. We are also exploring the possibility of connecting the environmental law enforcement and the management platform of public security punishment, which will be established soon, to enhance the deterrent effect. For example, if a cross-border heavy-duty truck was found of exceeding the emission standard without the capability of providing the evidence of rectification, we will notify the public security department to stop issuing the permit of accessing Beijing to the truck because of the misconduct. If any truck with the license plate of Beijing was found of exceeding the emission standard, the information will also go through the same system for a better supervision through tens of thousands of cameras, so that we can monitor and require an immediate rectification within the scope of laws and regulations. We can then easily and timely locate the vehicle through cameras, conduct on-site inspection, and penalize the vehicle strictly and seriously according to laws and regulations, which can ensure a deterrent effect to the vehicles with excessive emissions.

Fifthly, we need to focus on conservation. All heavy-duty vehicles are related to logistics, which means conservation is directly related to air pollution control. Currently, the population of Beijing is more than 20 million, the emission of the rigid demand of a safe city operation and the social production and livelihood of people is constantly taking a bigger proportion of the total pollutant emission. Hence, I would like to further call on your attention that green production from the society level and green lifestyle from the personal level are both closely related to environmental protection. Thank you!

The Economic Observer: Director-General Wen mentioned that coal firing is the priority of the air pollution treatment this year when he was introducing the situation of the comprehensive remediation of air pollution in winter. I would like to ask Director-General Wen that what are the impacts of coal firing on air pollution? What major measures will be taken by Tianjin in terms of coal firing remediation?

Wen Wurui: Thank you very much for your great question. Coal firing has a significant impact on the air quality in Tianjin. Tianjin is an industrial city in tradition and coal takes a huge part in its energy structure. In addition, heating during the winter is a rigid demand in Tianjin, which means coal is indeed the key task for the "five controls" of air pollution prevention and control. If we take a look at the contribution to $PM_{2.5}$ pollutions, coal firing is the biggest source, accounting for 25%. So coal firing pollution prevention and control is always the top priority.

We mainly adopted four measures to control coal firing.

First, coal volume reduction. Reducing the consumption volume of coal is the primary task. For years, we have been continuously adjusting the energy structure and strictly controlling coal consumption. In 2016, we reached the target of reducing 10 million tons of coal consumption stated in the *Action Plan on Air* by adopting these measures by one year in advance. On this basis, we have been further reducing coal consumption according to the requirement of the national stepped-up actions and the characteristics of Tianjin since the start of this year. This year, we shut down another 7 sets of coal-fired power units with the total capability of 860,000 kw. At the same time, according to the requirement of the stepped-up actions this year, we have conducted an all-around intensive inspection and found out 11,122 boilers of various kinds, including boilers for industrial use, heating, and commercial use. We only kept 184 boilers that reached the special emission limit or with ultra-low emission. The other 10,938 boilers were all shut down and remodeled to use natural gas, which was unprecedented. These tasks are all completed currently, which involved 4,630 industrial boilers, 2,284 for heating, and 4,024 commercial ones. Through this effort, the coal consumption in Tianjin can be reduced by 2.6 million tons. Through our years of effort, the proportion of coal in primary energy sources dropped from 54.4% in 2012 to 49% last year, expecting a further reduction this year; the consumption volume of natural gas was only 3.2 billion m^3 in 2012, but the number soared up to 7.4 billion last year.

Second, coal quality control. The quality of coal has a huge impact on air quality. The city government revised the quality standard of coal in Tianjin twice to constantly enhancing the requirement of coal quality. At the same time, we also published the government notice of promoting the management of coal businesses and operations to comprehensively enhance the management of the use of coal. It is currently the autumn and winter seasons in Tianjin, which is also the important time for the companies of coal operations, thermal power plants, heating businesses to purchase and store coal. We comprehensively strengthened the sampled inspection, supervision, and management of the coal quality in the city, dispatched on-site coal quality supervisors to the key companies in order to strictly inhibit inferior coal from getting into the market in Tianjin and ensure the coal quality during winter.

Third, strict emission standard. In 2016, 9 thermal power plants and 22 sets of coal-fired power unit in Tianjin all realized ultra-low emission. On this basis, the left 19 coal-fired power units in 4 self-operated power plants were remodeled to reach the ultra-low emission standard. Currently, all of the coal-fired power plants and self-operated thermal power plants can reach the emission standard. The 184 kept boilers I mentioned earlier can all reach the special emission limit or the ultra-low emission standard. Furthermore, we installed autonomous monitoring system for 24-7 supervision in 252 key companies related to coal operations in Tianjin, in order to ensure that their emission can meet the requirements.

Four, control over small-scale "scattered" coal burning. This is also the top priority of the key action this autumn and winter. I talked about the all-around intensive inspection in villages and households we conducted this year. In total, we found out 1.21 million households using bulk coal, in which 130,000 are urban residents, the rest are in rural areas. We consider the bulk coal control as a prioritized "hard bone" we need to resolve during this autumn and winter. The national requirement is to complete this mission in 3 years, but the Tianjin city shortened to 2 years. We have received strong support from the state through our efforts. Tianjin is now listed as one of the pilot cities of the clean heating program in Northern China. We adopted the following measures according to the principle of "choosing electricity or natural gas for heating based on the reality of different households" to treat the bulk coal issue in 610,000 households: firstly, to use electricity instead of coals wherever the national grid can support. For example, the 110,000 households in the coal-free zone of Wuqing are all using the most advanced air source heat pump. Secondly, to use natural gas instead of coals for heating wherever the natural gas pipeline or the LNG station is available. Thirdly, establish additional central heating supply systems wherever the heating pipeline can reach to realize centralized heating. The national stepped-up action plan required Tianjin to finish the remodeling in 290,000 households this year. We have completed 340,000 households, and will accumulatively complete 610,000 households by the end of this year, all of the rest will be remodeled before the heating season next year. I would like to especially point out that the 600,000 households that cannot be remodeled this year will use clean coals to reduce the pollutant emission as much as possible.

Generally speaking, it is fair to say that the coal control effort of the stepped-up action this year has the unprecedented scale, strength, and speed. This project is currently in full swing. That would be all from me. Thank you.

The Paper: My question is about supervision and accountability. I would like to ask Director-General Gao, how is the supervision and accountability situation in Hebei? You mentioned sets of statistics of accountability during your introduction. But what are the other measures you have taken other than the data?

Gao Jianmin: Thanks for the question. The measures for the air pollution control are quite clear, but it is still very hard to implement the best measures and realize the clearest goals if we do not define accountability. With the strengthening of the supervision of environmental protection from the central government, especially since the MEP strengthened its supervision and promoted information disclosure, many local governments were criticized through notices because of the weak implementation of treatment measures.

Through the promotion and deepening of environmental protection supervision in the past several years, issues like uncertain accountability, undefined responsibility, and solutions for the "last mile" problems are becoming much more significant in air pollution control. Hebei prioritized the implementation of haze treatment measures. Early this year, the Provincial Party Committee and the

Provincial Disciplinary Inspection Commission of Hebei required that weak treatment of haze should be listed as an important part of the "one accountability and eight eliminations" program. The first set of statistics I covered earlier was the detailed information of the "one accountability and eight eliminations" program in Hebei from the start of this year.

Since the kick-off of the stepped-up action in the autumn and winter, we established and published the *Temporary Regulation of the Quantitative Accountability of the Action Plan on Comprehensive Air Pollution Prevention and Control in the Autumn and Winter in 2017–2018* taking the quantitative accountability regulation of the MEP and the reality in Hebei into consideration, in order to establish a series of pertinent measures, to solve the issues of inadequate implementation and undefined responsibility. I would also like to introduce the detailed, practical, and quantitative accountability plan.

The first characteristic is that more attention was paid to the accountability of implementation and expanded the scope of accountability. We have prioritized the rectification of the issues discovered during the intensified inspection and follow-up inspection of the MEP as the focus of accountability. In addition, issues uncovered during the targeted inspection and law enforcement investigation in Hebei, misconduct reported by the public, matters assigned and supervised by the Provincial Environmental Protection Administration, and the air quality improvement during the 4th quarter of 2017 and the 1st quarter of 2018 are all included in the scope of quantitative accountability. At the same time, issues of human-induced interference to the operation of the air quality monitoring stations and falsified monitoring statistics are involved in the accountability scope, which spotlighted the result and target of the effort, expanded the accountability scope, and promoted the implementation.

The second characteristic is that we focused on the accountability of the responsible organizations to ensure the preciseness. Environmental protection is not only a task of the environmental protection departments, but it is also related to various aspects and sectors. So we expanded the subject of accountability horizontally and vertically. The horizontal expansion means that the subject of accountability is no longer limited to the leaders of the environmental protection administrations of cities and counties, it also includes related leaders and departments responsible for air pollution control, especially the departments responsible for the supervision of production capacity, environmental treatment measures of companies, dust control, vehicle emission standard, etc.

Vertically, it includes Party Secretaries and heads of towns, even the responsible person in villages and the person in charge of grid monitoring, which can efficiently solve the issue of "the last mile." Take the issue of crop straw burning as an example. Whenever such a case is discovered, not only the responsible heads of towns and villages should be held accountable, the relevant responsible departments should also be held accountable; meanwhile, so will the person directly in charge and the leaders responsible for such issues. If it is found out that such issues have happened in 3 or 5 towns in a county consecutively, the related departments and responsible leaders should be held accountable as well. If five authority units of the prefecture-level had such issues, the responsible leaders of the city-level should also be held accountable.

This is how we can truly establish an accountability system both horizontally and vertically, which covers all related responsible authority units, departments, and leaders.

The third characteristic is an innovation, which is that we have zero tolerance to the most important issues and focus on them for accountability. The MEP has conducted 14 rounds of intensified supervision, while Hebei Province conducted 2 rounds. We found out that some areas may have the same issues again and again, without the capability of learning from their past experiences for effective rectification and implementation. To solve such issues, on the one hand, we strengthened the supervision and inspection, asked the law enforcement department to conduct case filing, punishment, and rectification; on the other hand, we seriously held related parties accountable, had zero tolerance to the ones with repetitive misconduct and who refused rectification. For example, if we find out one case with unfulfilled rectification, we will hold the head of the town accountable; if there are two cases, we will hold the Secretary of the town's Party Committee and the responsible deputy head of the town accountable; if there are 4 or 6 cases, the head of the county and the Secretary of the county's Party Committee will be held accountable, respectively.

Another issue with the most concern from the public is the zero tolerance of human-induced interference on monitoring statistics. If there is even just one case discovered, the responsible deputy head of the county will be held accountable, two cases then the head of the county will be held accountable, and the Secretary of the county's Party Committee will be held accountable if a third case is discovered. Recently, with the consent from the provincial government, we kicked off the accountability procedure to 10 counties (cities, districts) with more flaws discovered during the rectification required by the MEP. Currently, these 10 counties (cities, districts) have held 152 responsible people accountable, which involved 50 persons at the section head level and above.

Through the innovation on those three aspects, Hebei enhanced the logic system of quantitative accountability, expanded the subjects of accountability, and seriously penalized the responsible persons, jointly formed an accountability system with strong determination and strict rules. By focusing on accountability, we effectively solved the issues of undefined responsibility of inadequately implementing the measures of the comprehensive air pollution control. It positively promoted the strict and precise implementation of the supervision and rectification responsibilities of the Party Committees and governmental departments at various levels. In the future, we will continuously uphold our principle and bravely shoulder our responsibilities to constantly conduct accountability measures and ensure their eternal effectiveness. Thank you.

Financial Times: I would like to ask Director-General Gao that some reports said that some factories were closed because of the air pollution control in the Beijing–Tianjin–Hebei Area, is that true? Another question is that what are the measures Hebei have taken to ensure the result of the treatment against haze? What are the expected effects? If the expectation cannot be reached, will there be any further enhancement?

Gao Jianmin: Thank you for the question. First thing first, it is not accurate to say that "factories were closed," or let's say it is still open to discussion. When it comes to the comprehensive air pollution control, the Northern China area has our specific climate condition and meteorological cycle. In autumn and winter, especially during the unfavorable meteorological condition for pollutant dispersion, if it was concluded that there will be some days with severe air pollution negatively affecting people's health after consultation on early warning and scientific analysis with the meteorological departments, we would have started the off-peak production plan according to the emergency response mechanism for heavy air pollution.

In order to proactively respond to the heavily polluted weather and reduce its harm to the public, we explored a set of effective measures during the past years, which is to establish early warning mechanism, off-peak production plan, and supervision implementation system to efficiently reduce pollutant emissions. In Hebei, different cities have completely different industrial structures and pollution sources. Instructed by the MEP, we conducted detailed and accurate basic analysis of the pollution sources of each city and developed lists of pollution sources. Especially for the industrial companies with relatively intensive emission, we conducted in-depth and scientific analysis on these companies' emission performance level, geographic locations, and their impact on the heavily polluted weather in the respective cities.

Based on scientific analysis, Hebei established off-peak production plan in heavily polluted weather in each city. The steel industry in Hebei generates serious pollution. In addition to promoting the pollution treatment facility, we also emphasized off-peak production in the steel industry. Key cities including Tangshan, Shijiazhuang, and Handan needed to reduce their production by more than 50%, the coking industry in the entire province needed to reduce its production by 30%. It is fair to say that we implemented "different policies for each plant" for the off-peak production plan. Plants could adopt off-peak production or rotational production whichever is more appropriate. We filed and penalized the plants exceeded emission standards and even instructed them to suspend production for rectification.

Off-peak production has been applied for years and some cities have developed effective mechanisms and measures. It has shown a positive effect on the reduction of pollutants. This year, we attached more importance to the scientism and operability when revising the response plan of heavily polluted weather, in order to reduce the social costs and public influence as much as possible. Take the companies and major projects currently under construction for an example, they can maintain the production normally as long as the pollution treatment facilities are in place and the emission can stably meet the standard.

The Beijing–Tianjin–Hebei area has a special geographical location and meteorological conditions during the heating season. If any unfavorable weather condition occurs, the environmental monitoring departments and meteorological departments will organize a scientific consultation. Once it is predicted to have heavily polluted weather, they will promptly release early warning and timely implement off-peak production according to the response plan, in order to promote the air quality and ensure people's health. Thank you.

Southern Metropolis Daily: My question is to Director-General Wen from Tianjin, which is about the treatment of the companies that "in unauthorized locations, lacked the proper certificates and failed to meet emissions standards." Statistics show that a total of 19,000 companies with such issues have been discovered and inspected in Tianjin. What is the current progress of the rectification? How hard is it? How will it impact the economic development of Tianjin?

Wen Wurui: Thank you very much for your question. This is a "hard bone" we must solve during the stepped-up action in autumn and winter that is extremely tough. From the start of this year, key responsible leaders of the Municipal Party Committee and the city government, including Secretary Li Hongzhong, personally instructed and continuously pushed forward this project according to the requirements made by the Party Central Committee and the State Council, as well as the guidance of Ministries like the MEP, which received significant achievements. We started with a comprehensive and intensive inspection of the companies with "location, certificate, and emission" issues, established a dynamic updating mechanism and a management book to keep the record. We found out nearly 19,000 companies with such issues and conducted treatment one by one based on the principle of "implementing measures according to various types, and different policies for each plant."

The first type is the companies with severe pollution and no possibility of a full remediation, which includes more than 9,000 companies. This type of companies usually has high energy and material consumption, intensive emission, and severe pollution that generate potential safety hazards, which is not in accordance with the urban planning, the requirement of land usage, or the principle of green development. This kind of companies was shut down based on laws and regulations. Currently, we have shut down 9,081 companies and implemented the measures of "cutting industrial water and electricity, and clean the material, product, and production equipment."

The second type of companies is the ones that meet related requirements with promising market prospects, upgrading conditions, and the possibility for on-site remediation. We are going to conduct on-site upgrading and remodeling to these companies. There are about 6,900 companies in this category. We developed remediation plans for upgrading and remodeling according to the principle of "different policies for each plant." After the joint review and approval of the related departments, including the environmental protection departments, the authorities of work safety supervision, and the administrations of water affairs, these companies could conduct remediation and restore production afterward. Now, a batch of such companies has completed the remediation and resumed production.

The third type is the companies with promising market prospects and favorable conditions for remediation, but are limited by their geographical locations or do not meet the planning requirements and the requests of land usage. About 2,900 companies fit in this category, and we relocated them into industrial parks for further remediation. This task will be finished next year; work on the second type that requests on-site remediation will be completed by the end of this year.

Another type is the traditional clusters of companies. For example, the Caozili Town of the Wuqing District used to be the hometown of silk flowers with hundreds of years of history. There are about 500 companies with more than 10,000 people manually producing silk flowers. For this type, we established special plans for an overall improvement, including unified water supply, drainage, and governance, which can realize upgrading and transformation while protecting the traditional processes.

You asked if there has been any difficulties and economic impacts. Frankly speaking, in the short-term, there are some difficulties and impacts from the company level. From the macro perspective, difficulties and impacts are only partial for the long run, which have to be overcome in order to realize green and sustainable development. In this process, we profoundly realized that the remediation of the companies with "location, certificate, and emission" issues is necessary to boost green development. Although it is significantly tough and difficult to shut down and upgrade so many companies in such a short period of time, it is something imperative. As some of these companies were developed in the 1990s or even 1980s, production equipment can be extremely outdated with high energy consumption and weak company management capability. As a matter of fact, many companies are struggling and unsustainable. Therefore, we need to promote and accelerate the remediation and upgrading of these companies. However, there are some difficulties during this process. Just as General Secretary Xi Jinping said: "promoting the green development and lifestyle is a profound revolution of the outlook on development." So the Municipal Party Committee and city government of Tianjin emphasized its importance from the top-level, printed, and distributed dedicated document to push forward the implementation. Leaders including the Secretary of the Municipal Party Committee and the Mayor conducted thorough investigation personally and shouldered the responsibility of ensuring the remediation of different districts, respectively, to promote the upgrading and transformation of these companies one district after another and the overall air pollution prevention and control. So I truly think the treatment of the companies with "location, certificate, and emission" issues is needed for the green development.

Secondly, I would like to talk about if there is any economic impact. I just said that there have to be some impacts on some areas and companies. But if we take a look at the big picture for the long-term, it is like vacating the cage for new birds. The Beichen District of Tianjin is located in the suburban area, where is packed with the companies to be shut down because of "location, certificate, and emission" issues. A total of 1,600 companies were closed, which took 3,000 mu. On average, a company took more than 2 mu and contributed 30,000 RMB of taxation, which means the outcome of every mu of land is less than 20,000 RMB. After closing those companies, there are more lands, resources, and environmental spaces in the Beichen District for further development and the optimization of industrial structure. It is also beneficial for us to solve some issues with potential security concerns and further realize green development.

My third point is that we received strong support of the public during the treatment of the companies with "location, certificate, and emission" issues. Between this April and May, the Environmental Protection Inspectors from the Central Government stationed in Tianjin, received more than 4,000 petition letters and reports from the public, among which a large proportion was about this kind of companies. It shows that the public is very dissatisfied with and concerned about these companies. Therefore, according to the requirement of "do whatever is requested by the public" stated by General Secretary Xi Jinping, we must take actions to strengthen the governance. Though the treatment of the companies with "location, certificate, and emission" issues affects a small portion of people, generally speaking, it is beneficial to the major public and the big picture.

Meanwhile, we felt the support from most companies as well. Without their support, we could not have conducted remediation to 19,000 companies and closed more than 9,000. I have a story to share. June was the most critical moment of shutting down the companies with "location, certificate, and emission" issues. The Cycling Association of Tianjin led dozens of companies to the Tianjin Environmental Protection Bureau for a discussion to solve this issue. After the discussion, responsible persons of two companies said that they had been prepared for a fight and came in a sulk. But after learning about the series of policies of combining the unblocking and implementation of reforms, providing classified guidance, and implementing categorized policies deployed by the Municipal Party Committee and the city government to finally realize upgrading, transformation, green development, and industrial development, they understood the next step and were no longer angry. During the discussion, I talked about how hundreds of small plants were integrated and founded the Shanghai Phoenix Bicycle Company in 1958, which realized industrial and scale development, eventually built up a renowned brand. We suggested that they should follow the same development model. Currently, the bicycle industry in Tianjin is thoroughly implementing the remediation requirements of the companies with "location, certificate, and emission" issues. We are confident and determined to complete the comprehensive remediation of these companies as scheduled to realize a better and green development. That is all from me. Thank you.

People's Daily: I have a question to Director-General Fang. How can Beijing achieve the goal of keeping the $PM_{2.5}$ below 60 $\mu g/m^3$?

Fang Li: Everyone is very concerned about air quality. As a matter of fact, people's understanding of air quality has become more scientific and comprehensive. Air quality is the final reflection of the discharge of social production and the emission of human activities to the nature overlaying with the weather condition. Therefore, the air pollution control depends on the effort of us human.

When it comes to the evaluation of air quality, if you are a sports fan, you may understand that the most significant ambiguity always arises from the ones scored by judges. The highest and lowest 2 scores are usually removed before calculating the final score. The weather condition I mentioned may vary every year. It is not reasonable to simply compare the air quality of today to that on the same day of the last year. However, it is more scientific if the term of a comparison is longer, for example, a comparison of the data for one year is more accurate than that of one day; a comparison of the data for three years is even better. If we use some methods of scientific analysis from the sociology and statistics fields, we will have some useful conclusions. We are used to evaluating the air quality with the cycle of one year, which is an approach. However, if we take it one step forward, remove the 10% days with the best and worst air quality and only focus on the change of the 80% in the middle, the evaluation will be more objective with relatively less influence from the weather condition.

You asked about how to achieve the goal of 60 $\mu g/m^3$ in the final year of the *Action Plan on Air*. I think we have partly answered the question today, and statistics also showed some evidence. Firstly, this press conference is held in the MEP, which represented the solid determination nationwide.

Secondly, heads from the environmental protection authorities of Beijing, Tianjin, and Hebei are sitting here together, which showcases the strong coordination and effective cooperation of the joint prevention and control of the three areas. You may have also noticed the support and effort from the entire society through our earlier introduction.

Thirdly, statistics speak louder than words. Many of you have been participating in this press conference from the start of this year, who may have a clear understanding of the data this year. In this January and February, the average $PM_{2.5}$ in Beijing rebounded more than 60% compared with the same period of last year. We did some analysis immediately and found out that if we wanted to reach the goal of 60 $\mu g/m^3$, the $PM_{2.5}$ for the next ten months need to reach the best level in the same period every month. The monitoring data became available 4 years ago, which means we have statistics of 2013, 2014, 2015, and 2016. You can see from the data that the annual average $PM_{2.5}$ has been decreasing every year, but it did not necessarily mean that the best level of $PM_{2.5}$ in a certain month always occurs in the latest year. It is determined by the weather condition of the particular month. It proves that the air quality is highly related to meteorological conditions, which is definitely the truth for the current phase. Data from the past 4 years shows that the emission has been decreasing and the air quality has been constantly improving through the effort of our entire society.

From this March, except September, 7 in 8 months reached the lowest level of $PM_{2.5}$. It means that although the meteorological condition had its impact, the overall emission in this area still decreased via the unified guidance of the MEP and the joint prevention and control between regions. Compared with the same period of 2013 when the average $PM_{2.5}$ of Beijing was 89.5 $\mu g/m^3$, the goal of 60 $\mu g/m^3$ means that we need to realize a reduction of 1/3 this year. As of yesterday, the $PM_{2.5}$ reduced by 34% over the same period of last year. The human effort will become even more apparent in the last two months of this year. On the one hand, we should spare no effort to fully implement all measures. The public will notice our effort as well. Execution and actions of the government and society are more important than the goal itself.

The confidence of the air pollution prevention and control in Beijing is not only from the aforementioned aspects, it is also based on the background: the newly disclosed general planning of Beijing; and the other 2 major tasks mentioned in the report of the 19th National Congress of the CPC, namely, the Winter Olympics and relieving the noncapital functions from Beijing. If you have paid attention to the general planning, you may notice that Beijing will adopt a development model with reduced resources consumption. The construction land area in the urban and rural area will be reduced from the current more than 2,900–2,860 km^2 in 2020.

Take the treatment of companies with "location, certificate, and emission" issues for an example. You can go and check out the Daxing District, where used to have a number of industrial parks with such companies. After implementing the remediation and relocation measures, some parts became eco-parks, other lands were transferred to high-tech companies for a better development of this area. You should also have noticed the remediation of small streets and narrow alleys in the center-city area, which promoted the overall environment. In the future, we need to clean up the entire city, which is also stated in the new general planning. Considering the background of moving out the non-capital functions, ameliorating the environment, including the air quality, is not only a task for the environmental protection department, but also a requirement to every industry and department in Beijing. We are very confident to the air quality in the future, because that we can improve the environmental quality and the overall condition of the city through our effort of building up a world-class harmonious and livable city, in accordance with the requirements of the central government. Thank you!

People's Daily: Hebei and Tianjin have reached the milestones for the current phase. Will there be any further rigid goals?

Gao Jianmin: The completion of the tasks stated in the *Action Plan on Air* does not mean that the pressure on us is less, it became even bigger.

Firstly, as we all know, the adjustment of industrial structure cannot be done overnight. It is true that we have reached certain milestones of air pollution control, but there is still a huge gap between the air quality and public expectation. Their expectation is our greatest pressure.

Secondly, there is still a long way to go for the green development and a fundamental improvement of the air quality in Hebei. We have to change the outdated industrial and energy structure, strengthen air pollution control, and consolidate the results, in order to continuously improve the air quality.

Thirdly, as a big area, Beijing, Tianjin, and Hebei need to conduct joint coordination and control. Many standards in Hebei need to benchmark and be aligned with those in Beijing and Tianjin to promote the environmental governance of Hebei to a higher level.

Fourthly, General Secretary Xi Jinping requested Hebei to focus on air pollution control. We need to further adapt and develop the existing effective measures and firmly improve the air quality, in order to implement General Secretary's requirements and solve the issue with the most concern. We will work harder to make our skies blue again and bring the happiness of enjoying the blue sky and white clouds more often to our people. Thank you.

Wen Wurui: I would like to briefly add something. Last year, the average $PM_{2.5}$ of Tianjin was 69 $\mu g/m^3$, which was 28.1% lower than in 2013. We completed the task of the *Action Plan on Air* ahead of schedule, but it is still highly exceeding the national standard of 35 μg/m^3. Therefore, we just took the first step in air pollution control, the prominent issue of air pollution during autumn and winter remains to be well resolved. Statistics show that in the past years, the reduction of $PM_{2.5}$ in autumn and winter was not obvious. It means that the air pollution control issues during autumn and winter, especially the response to heavily polluted weathers, are still the most essential ones. To this end, the stepped-up actions against air pollution this autumn and winter gave us higher goals and stricter requirements. For Tianjin, a 25% reduction of the $PM_{2.5}$ during autumn and winter is a very high target. From last October to this March, a 25% reduction means that the $PM_{2.5}$ need to drop from 95 μg/m^3 to about 70–71 μg/m^3; while the requirement of reducing the number of days with heavy pollution by 20% is another tough requirement. During the last autumn and winter, there were 35 days with heavy pollution in Tianjin.

To reach this target, I mentioned that we have established plans for the stepped-up action and adopted 10 measures for implementation. The strength and intensity of emission reduction are both unprecedented. We have taken measures to treat both the "symptom" and the "root cause", as well as some newly adopted ones, such as off-peak production, construction, and transportation. I believe these measures will definitely be effective.

Director-General Fang Li said that our hard work also needs the help of the weather. It is indeed the case. Since the comprehensive implementation of the stepped-up action, between October 1 and 25, the $PM_{2.5}$ in Tianjin was 54 μg/m^3, 24% lower than the same period of last year, which was also the historical best. We achieved unprecedented results. However, during a large-scale regional heavy pollution process between October 26 and 28, in 3 days, the $PM_{2.5}$ of the month increased by 10 μg/m^3. This heavy pollution process dragged the improvement rate of the monthly $PM_{2.5}$ in Tianjin down from 24 to 6.1%. In Tianjin, the coastal area was more significantly influenced by the heavy pollution compared with the Northwestern area. On October 27, the $PM_{2.5}$ of the Binhai New District in the east reached 221 μg/m^3. On the contrary, Baodi and Wuqing Districts in the north had a $PM_{2.5}$ of about 160 μg/m^3.

It shows that heavily polluted weather is the most prominent and tough issue influencing the air quality in autumn and winter. Key actions during autumn and winter should be taken against it. We need to fully strengthen consultation on early warning as well as emergency response and joint prevention and control, to minimize the impact and shorten the duration of pollution, which are the core of ensuring the air quality in autumn and winter.

You talked about stop production, which I would touch upon further. From October 1, Tianjin established measures to stop the earth and stonework of construction in the established city area. Since Tianjin is currently in the stage of fast and large-scale development and construction, with more than 2,500 construction sites in the city. Recent years, Tianjin decently managed construction sites, road cleaning, and crop straw burning, which controlled dust pollution. But during winter, because of the numerous construction sites and dry weather, dust pollution is more often, which contributes to 18.2% of the $PM_{2.5}$ in winter. We adopted targeted controlling measures to solve this issue.

First, suspended construction based on categories. Instead of suspending all construction sites, only the ones related to earth and stonework, construction wastes transportation, and accepting and discharging materials are suspended. Other construction work can still be carried out as usual without any impact. However, we adopted strict control of the processes causing dust pollution.

Second, suspended construction based on locations. We mainly suspended the construction in the established areas of cities because it has the greatest population density that may lead to the biggest impact on the public. In addition, suspending the earth and stonework can stop the shipment of construction wastes that could generate dust, which can, in turn, reduce the discharge of relevant pollutants.

Third, suspended construction based on scenarios. The construction of the major projects related to people's well-being, for example, railways and forestation projects, may need to be done in specific seasons. We gave these projects with special permissions to continue with the earth and stonework during this season. The companies need to apply for the permission. After the review of industrial departments and the approval from the city government, the permission will be printed and published on the construction site for public supervision. The construction sites with special permissions need to ensure 6 "100%" to effectively control the dust pollution. Construction companies support these measures and have made preparations in advance to ensure a solid implementation.

Economic Daily: During the two sessions, Premier Li Keqiang said that the scientists who can "solve the tough issue of haze and smog" should be amply rewarded. The MEP launched the project of analyzing the cause and control of heavy air pollution. When can the public know the exact cause of haze and smog? When will the problem be solved?

Liu Youbin: I would like to briefly answer the question. In September 2013, the State Council issued the *Air Pollution Prevention and Control Action Plan*, or the *Action Plan on Air* in short. It was released after extensive expert review and fully hearing from their opinions. Since the implementation of the action plan, experts actively participated in relevant work with the support of many national science and technology projects, which ensured positive improvement of the basic research and technology R&D of the air pollution prevention and control in China. For example, we came up with the framework of the formation mechanism of regional combined air pollution, independently researched and developed the technologies of dedusting the coal firing smoke, desulfurization, and denitration. Now the ultra-low emission technology of the electric power industry is at the leading position of the world. We realized detailed forecast of 3 days and trend forecast of 7 days. We also established dispatch model for regional joint prevention and control of the emission reduction of major pollutants, promoted the standard system of air quality and pollutant emission, which strongly support the improvement of air quality in key areas and cities.

Of course, we need to be aware of that there is no end for scientific research. Along with the constant in-depth development of air pollution prevention and control, there have been new requirements for the science and technology work, which brought new situations and issues to us continuously. Other director-generals mentioned that one of the current problems is the frequent regional heavy pollution in autumn and winter with $PM_{2.5}$ as the major pollutant, which raised higher and more urgent requirements to the scientific support of a more refined pollution control.

The 170th Executive Meeting of the State Council instructed the MEP to take the lead, with the joint effort of science and research resources from various departments, including the agriculture, industry and information technology, public health, meteorology, the CAS, and universities, to conduct focused research on the cause and control of heavy air pollution in autumn and winter in Beijing–Tianjin–Hebei region and periphery. Currently, based on the existing achievements of researches and studies, the MEP and related department have jointly established the *Implementation Plan of Research Program on the Cause and Control of Heavy Air Pollution* in accordance with the principle of "taking measures based on problems and requirements," targeting at making scientific decisions and implementing precise policies to treat the heavy air pollution in Beijing–Tianjin–Hebei region and its periphery in autumn and winter. We established 4 special projects and 28 specific topics related to the cause and source of the heavy air pollution in Beijing–Tianjin–Hebei region and its periphery, the emission status evaluation and the advancement of prevention and control technologies, the comprehensive scientific support of the decisions on air pollution prevention and control, and the impact of air pollution on people's health. Recently, Minister Li Ganjie announced the start of the stepped-up action.

This project drew on the "atomic and hydrogen bombs and man-made satellite" research model, gathered all outstanding members and scientists, and built up a research team of the national association of air pollution prevention and control with more than 1,000 people. We adopted the research tracking mechanism of "fixing the output quota for each household," instructed the 28 expert teams for on-site investigation and guidance in the "2+26" cities, started the research of the scientific support of "different policies for each city," and established a shared management platform. Currently, the 3-dimensional comprehensive monitoring measure has been initially established, which ensured the fundamental capability of heavy pollution forecasting, monitoring, and the fast analysis of its cause.

You particularly asked about when will we know the cause of haze. Other director-generals also said that different places have different industrial and energy structures, as well as various meteorological conditions, which means the composition and source of $PM_{2.5}$ are different. There are differences between Beijing, Tianjin, and Hebei, cities in Hebei may also have various statuses, so it is hard for us to identify a universal cause of $PM_{2.5}$ that is applicable to all places. Therefore, the experts were deeply involved in the work of these cities and counties, actively monitored, analyzed, and studied the composition, to provide more targeted prevention measures of air pollution control and scientific decisions to the emergency response plan of heavy air pollution in different places. I believe that as the experts continuously develop their studies, we will better understand the precise source of

$PM_{2.5}$ in different areas. Our team will also timely disclose related information to the public. Thank you.

Reuters: You all mentioned the enhanced effort of remodeling coal-fired boilers in different places. But we also noticed that the price of natural gas rose dramatically recently. We know that some companies and plants in the Beijing–Tianjin–Hebei area can no longer afford the cost of natural gas after the "switch from coal to gas" and are more concerned about the natural gas supply in the future. I would like to know what is your perspective on this issue? What's for the next step?

Liu Youbin: The "switch from coal to gas" is an essential measure for the air pollution control in autumn and winter. We know that one of the major causes of air pollution is the coal-based energy structure. In order to promote air pollution prevention and control, we must constantly adjust the energy structure, which is also a common recognition of the international society. After rounds of review by experts, the "switch from coal to gas" has been included in the work plan in many areas and become a very important measure for air pollution control. Now, regional governments are pushing this project forward according to the plan.

The natural gas supply issue you just mentioned is a topic currently under discussion of the MEP and other related ministries and departments. Through our communication and exchanges, we will fulfill the requirement of clean energy for heating in autumn and winter. Even if there is a shortage of natural gas supply for civil use in some places, we will make a timely adjustment to ensure the resources. As for the price problem you talked about, there have been some price fluctuations in some areas. As far as I know, the responsible people from the MEP actively communicated with related departments to ensure a stable market and a smooth and orderly implementation.

Wen Wurui: I mentioned that there are 1.21 million households that need to switch to natural gas or electricity, in which 600,000 will wait until the next year. The reason for us to wait till the next year is that electricity is more secure and cleaner, so that we want more households to switch to electricity, as many as possible. We focused more on "switch from coal to electricity" during the transformation, actively promoted the grid construction to catch up with the schedule.

The situation of natural gas supply is quite the case introduced by Director-General Liu. If there is a shortage of natural gas supply, especially during the peak hours, we are going to adopt some measures, like adjusting the non-civil use natural gas to residents, according to relevant plans of the state and Tianjin city, to solve the problem.

Gao Jianmin: I would like to add something. For the established city areas and surrounding rural areas in Hebei, there are some places need to make the adjustment and switch to natural gas or electricity, whichever is suitable, to complete the remodeling of heating with clean energy. Hebei is an important part of the pilot project of heating with clean energy in Northern China. Based on our current understanding, eliminating the use of bulk coal is the priority of solving the emission problem of heating during autumn and winter in the rural area, which is essential for making our skies blue again.

We have made some progress in Hebei. All cities and counties have been actively working on the plan of remodeling 1.8 million households we established early this year. With the support and recognition of the public, we may go beyond the target. Now, all 1.8 million households have completed the preparation work outdoor and are ready for the remodeling. In order to ensure the natural gas supply, Hebei actively communicated with suppliers like PetroChina, Sinopec, and CNOOC, to expand the source of supply, promote the construction of storage facilities, and establish detailed and applicable emergency response plan with the strong support of the MEP. Currently, we have ensured the natural gas resource of 6.5 billion cubic meters for the heating season, which can guarantee the gas supply for the households that switched coal to natural gas as required by the province. Thank you.

Liu Youbin: That is the end of today's press conference. Thank you for your participation!

Regular Press Conference (November)

November 23, 2017

On the morning of November 23, the MEP held the regular press conference for the month. You Yanxin, Counsel (Director-General level) and Deputy Director General of the Department of Planning and Finance of MEP, and Wang Jinnan, President of the Chinese Academy of Environmental Planning of MEP, introduced the planning and finance work related to environmental protection. Liu Youbin, Counsel (Director-General level) of the Department of Communications and Education of MEP, reported progress in recent key tasks of environmental protection. They also answered questions from the journalists present.

Ministry of Environmental Protection of the People's Republic of China,
2017 Press Conference Records of Ministry of Environmental Protection, the People's Republic of China, https://doi.org/10.1007/978-981-13-7330-5_12

MEP November Regular Press Conference (1)

MEP November Regular Press Conference (2)

Liu Youbin (Moderator): Good morning, friends from the press. Welcome to MEP's regular press conference of November.

In his report delivered at the 19th National Congress of CPC, President Xi Jinping drew the blueprint for building Beautiful China. Good planning and finance work is a major support and safeguard to strengthen environmental protection, progressively advance ecological civilization and build Beautiful China.

Today we have with us You Yanxin, Deputy Director General of the Department of Planning and Finance of MEP, and Wang Jinnan, President of the Chinese

Academy for Environmental Planning of MEP. They are about to brief you about environmental protection planning and answer your questions.

First, let me briefly introduce MEP's progress in recent key tasks.

1. Special action for law enforcement for protecting the drinking water sources of the Yangtze River Economic Belt

To implement General Secretary Xi Jinping's instructions on well-coordinated environmental conservation and avoiding excessive development in the Yangtze River Economic Belt, follow the strategies made by the CPC Central Committee and the State Council and fully implement the *Outline on Developing the Yangtze Economic Belt*, MEP, since 2016, has been carrying out targeted law enforcement to protect centralized drinking water sources in cities at and above the prefecture level in the Yangtze River Economic Belt, combating illegal activities in protection zones and urging localities to fulfill their responsibilities.

The special action is mainly about two tasks. The first is to check the implementation of the zoning system for protecting drinking water sources. The key lies in ensuring that protection zones are delineated and equipped with conspicuous geographical landmarks and warning signs at the boundary according to the law. The second is to address environmental illegalities in protection zones, which involves banning illegal sewage outlets and forbidding or shutting down illegal construction projects.

Over the past year, MEP has strengthened guidance and on-site supervision to push localities concerned to step up rectification. This helps resolve some major problems. Drinking water source protection zones have all been delineated, and environmental illegalities have been basically identified. Breakthroughs have been achieved in resolving legacy problems. For example, oil ports in some protection zones in Huangshi of Hubei Province, Yiyang of Hunan Province and Wuhu of Anhui Province, which used to cause persistent and difficult problems, have been addressed.

By November 13, 2017, altogether 465 problems have been resolved, accounting for 95% of the total. In the 126 prefecture-level cities in the Yangtze River Economic Belt, 110 have completed their rectification tasks. Of the 25 problems yet to be resolved, most are persistent and difficult legacy problems, namely, ten illegal buildings and industrial enterprises in protection zones, nine rural nonpoint sources and six illegal sewage outlets. Considering that these problems concern the interests of multiple parties and involve many functional departments, local governments need to take the lead in addressing them. In addition, environmental protection and supervision on law enforcement in the protection zones involve multiple departments such as environmental protection, water conservancy, agriculture (fishery) and transportation. Therefore, while each of them should better fulfill their own duties and responsibilities, they should also strengthen coordination and interaction to jointly protect the environment.

Next, MEP will carry out vigorous inspections of centralized drinking water sources in cities at and above the prefecture level in the Yangtze River Economic Belt. The five-step approach of inspection, assignment, follow-up inspection, admonitory talks, and targeted inspection[1] will be adopted to keep track, timely reveal relevant information to the public and subject itself to the supervision of the public.

2. "Green Shield 2017" targeted action for supervising and inspecting national nature reserves

To implement the *Circular of the General Offices of the CPC Central Committee and the State Council on the Inspection of and Solutions to the Ecological and Environmental Problems in Gansu Qilian Mountain National Nature Reserve and the Lessons Learned*, seven ministries including MEP formed ten national follow-up inspection groups dedicated to the "Green Shield 2017" targeted action for supervising and inspecting national nature reserves. The follow-up inspection started in October 2017, and up till now the ten follow-up inspection groups have covered altogether 112 national nature reserves in 31 provinces (autonomous regions and municipalities).

Overall, all localities value the "Green Shield" targeted action. Party committee and government leaders of more than ten provinces (autonomous regions and municipalities) including Hebei, Yunnan, and Tibet have held meetings to assign tasks and to more vigorously rectify violations against laws and regulations. The vast majority of localities have established a multi-department work mechanism, developed a management logbook targeting violations against laws and regulations and carried out inspections in protection zones including self-checks and spot checks by provincial teams. According to rough statistics, localities have processed leads on over 18,800 problems involving nature reserves and produced opinions on the handling and rectification of all of them. Most problems have been addressed and ecological restoration measures are being implemented. Accountability-wise, localities have held over 960 people accountable and punished 49 director general level and over 210 division head level officials, deterring, warning and educating other officials.

The follow-up inspection revealed that some localities had not fully grasped the necessity of nature reserve protection. They are not thorough or firm in investigating and penalizing violations against laws and regulations, holding the personnel responsible accountable and imposing pressure on lower level governments. A small share of localities are slow in rectification. In addition, most nature reserves share some common problems such as the lack or absence of noticeable warning signs, insufficient publicity, and weak and ineffective management and infrastructure maintenance. The follow-up inspection groups came up with rectification requirements and suggestions on the spot.

Next, MEP will analyze and draw lessons and experience from the follow-up inspection. We will conduct admonitory talks with local governments and nature

[1]First, central inspectors spot a problem, assign the local government to address it, then follow-up inspections will be conducted to see to the resolution of the problem. MEP will carry out admonitory talks with local governments that are slow in rectification, and conduct targeted inspection of those that have major problems but take no rectification action.

reserve management agencies who are not rigorous in their violation investigation and punishment or forceful in rectification, and urge them to speed up rectification to ensure that the targeted action is effective.

3. Six regional supervision centers renamed as inspection bureaus

Recently, the State Commission Office for Public Sector Reform has approved MEP's application to change the six environmental supervision centers in northern, eastern, southern, northwestern, southwestern and northeastern China from public institutions to dispatched offices and rename them as inspection bureaus.

Currently, MEP has distributed to the six bureaus the "three-specification" plan that specifies their responsibilities, internal agencies, and staffing. Bureau director generals were appointed and the bureaus are running in a coordinated and orderly manner.

The regional supervision centers were initiated as pilots and operated as fully fledged agencies in 2002 and 2008 respectively. They have played a positive role in strengthening environmental protection law enforcement, responding to environmental emergencies and mediating interprovincial pollution disputes. However, the supervision centers were handicapped by their nature as public institutions and not authoritative in supervision and law enforcement. This status and mandate issue is resolved by their transition from public institution to dispatched office, an administrative agency. This move will effectively promote the modernization of China's national environmental governance system and capacity.

The six regional bureaus will further hold government officials concerned accountable and work with the National Environmental Protection Inspection Office to build a national accountability system for environmental protection matters, improve the inspection mechanism and provide a strong safeguard for the central environmental inspection.

Now let me give the floor to Ms. You Yanxin.

You Yanxin: Good morning, friends from the press. I am delighted to have this chance to talk to you. First, let me thank you for your long-time support to our planning and finance work.

The planning and finance work serves as a guide, support and safeguard to environmental protection effort. According to MEP's "three-specification" plan, the Department of Planning and Finance is mainly responsible for organizing the compilation, evaluation, and assessment of comprehensive environmental protection plans; comprehensively analyzing the situation of environmental economy and serving as the liaison agency for such special programs as poverty reduction and pairing assistance for the development of western China; coordinating fixed-asset investment direction, scale and channels in areas of environmental protection and working with relevant departments for organization and implementation matters; drawing up ecological compensation policies and organizing their implementation; undertaking the implementation of permits for pollutants discharge, total pollutant emission control and permit trading; drafting and supervising the implementation of regulations and rules on management of budget, finance and state-owned assets, government procurement and internal audit; supervising, inspecting and managing the performance of projects sponsored by special funds from state revenue.

You Yanxin, Counsel (Director-General level) and Deputy Director General of the Department of Planning and Finance of MEP

This year, departments of environmental protection planning and finance nationwide have been earnestly implementing decisions made by the CPC Central Committee and the State Council. They follow the "strict, real, thorough, effective and fast" principle,[2] focusing on improving environmental quality while considering other factors holistically. Progress has been secured in all aspects. This year MEP has been working harder on the following five tasks and playing a stronger role, which is different in each task:

First, formulating and implementing a series of key plans and playing a stronger leadership role. MEP has mainly completed four tasks. (1) It facilitates the implementation of the 13th Five-Year eco-environmental protection plan, which is ahead of schedule now. (2) It has issued and distributed the eco-environmental protection plan of Yangtze River Economic Belt, a move that translates the principle of "well coordinated environmental conservation and avoiding excessive development" into a road map and construction plan. (3) It has formulated an eco-environmental protection plan of Xiong'an New Area to make it green and livable. It is committed to eco-environmental protection on all fronts in Xiong'an and has initiated a series of pollution treatment programs. On November 13, MEP signed a strategic cooperation agreement with the people's government of Hebei Province, which was widely covered by the media. (4) In line with the principles of

[2]Having strict requirements on assessment and quality, discovering real problems, implementing thorough management, doing effective work, and making fast progress.

the 19th CPC National Congress and the timeline for building a great modern socialist country in all aspects, MEP has launched strategic research on medium- and long-term eco-environmental protection.

Second, stepping up situation assessment and analysis as well as regional coordination and playing a stronger decision-making coordinator role. In this regard, MEP has mainly done four things. (1) MEP strives to better inform decision-making as it makes more in-depth and comprehensive analysis and assessment of environmental economy situation, keeps track of the progress and makes known new developments, proposes solutions and establishes a comprehensive decision-making mechanism for the environmental economy. (2) MEP holistically manages environmental protection binding indicators; it follows the principle of viewing environmental quality as the core and prioritizes it over quantity; it has established and improved a plan management and assessment mechanism by combining all kinds of assessments into one. (3) MEP has, as per Article 27 of the latest *Environmental Protection Law*, put in place an annual mechanism for reporting on environmental situation and goal status. (4) MEP has been implementing the strategy of coordinating regional development by addressing difficulties in poverty alleviation and facilitating pairing assistance.

Third, optimizing investment scale and structure and playing a stronger supporting role. Since the 18th CPC National Congress, the central government has been increasing funding for environmental protection. In 2017, special environmental funds earmarked from state revenue will reach 49.7 billion yuan. Many key projects have been implemented for water, air and soil pollution prevention and control, rural environment improvement, ecological protection, and restoration of mountains, rivers, forests, farmlands, lakes and grasslands and capacity building. All these efforts play an important role in improving the environment. At the same time, we pay much attention to early-stage preparation for environmental protection investment projects, consolidate the project reserve and provide technical guidance and training for localities one by one.

Fourth, managing environmental protection projects supported by special funds more rigorously and playing a stronger regulating and supervising role. To better regulate funding management, reduce interference from human factors and make funding allocation fairer and funding use more efficient, MEP adopted three measures. First, it has innovatively used the factor analysis method[3] to earmark special funds and tied fund appropriation to project performance and budget execution. Second, it has stepped up supervision and inspection, appropriated special funds by the season and put in place performance assessment pilots. Third, it has strengthened auditing supervision. Referring to the mode used in the National Audit Office, MEP supervised and inspected projects supported by special funds and tightened up financial discipline.

Fifth, innovating in environmental control policies and mechanisms and playing a stronger efficiency-boosting role. To that end, MEP has mainly been working on

[3]Decision-making based on a set of predetermined factors, the weight of which can be adjusted.

three tasks. First, it seeks to improve the ecological protection compensation mechanism by launching the upstream–downstream horizontal[4] eco-compensation pilots for Xin'an River between Zhejiang and Anhui Province, Jiuzhou River between Guangdong and Guangxi, Dongjiang River between Jiangxi and Guangdong, Tingjiang–Hanjiang River (Tingjiang River is in Fujian Province and flows into Hanjiang River in Guangdong Province) and Luanhe River (a river in Hebei Province, whose water is diverted to Tianjin Municipality) and by expanding the pilot scope. Second, it improves the market-based mechanism by promoting third-party environmental control and permit trading. Third, it pushes forward the PPP model.

Besides the abovementioned five tasks, we have also been carrying out another important task, permit for pollutants discharge, which we will brief you about separately. You can also visit our website or follow our official WeChat public account for more information.

The Department of Planning and Finance works on a wide range of tasks. It undertakes more and more general tasks particularly since new departments were established for water, air, and soil, respectively. Overall, all tasks are being carried out in an orderly, robust and productive manner. Our planning and finance work has come to a new historical point. As the relevant background documents have been given to you, I will not elaborate on that. Thank you.

Liu Youbin: The floor is open now.

Beijing Evening News: People always pay much attention to the emission of big diesel-fueled trucks in the Beijing–Tianjin–Hebei region and the ensuing pollution. Freight in the Beijing–Tianjin–Hebei region mainly relies on expressways and thus diesel-fueled trucks are major polluters. It is recently reported that MEP has conducted research and proposed increasing the share of railway in freight in the region. Can you elaborate on it please? Thank you.

Liu Youbin: To eliminate heavy air pollution, we have been delving into its causes. Experts use a very revealing metaphor to illustrate them, saying that the heavy-industry-dominated industrial structure, the coal-dominated energy mix, and the expressway-dominated freight transport structure are the "three mountains" that keep air quality from improving. In recent years, the air quality in the Beijing–Tianjin–Hebei region has been improving overall. However, the concentration of NO_2 does not drop as that of $PM_{2.5}$, PM_{10} and SO_2 did; rather, it still exceeds the standard, except in some cities. In the first of 2017, the average concentration of NO_2 is rising. In recent pollution processes affecting Beijing, the share of nitrate in $PM_{2.5}$ is the largest and rises the fastest.

[4]"Horizontal" compensation means compensation between local governments at the same level in the river basin concerned, vis a vis the vertical compensation that comes from governments at higher levels.

According to the research of experts, the energy consumption per unit of ton-mileage and the emission of major pollutants per unit of freight volume in railway freight are only 1/7 and 1/13 of that in road freight respectively. Thus, channeling freight from road to railway is of great significance to conserve energy, protect the environment and improve air quality. According to statistics, road freight accounted for 84.4% of the freight volume in 2016 in the Beijing–Tianjin–Hebei region, and most of vehicles used were diesel-fueled heavy-duty vehicles. Heavy-duty vehicles in use number 830,000, accounting for 4% of vehicles in use, and their nitric oxide emission accounts for 1/5 in the region. This year, the number of heavy-duty vehicles in use in the Beijing–Tianjin–Hebei region has been increasing at double digit. With their huge number, fast growth, and high emission of nitric oxide, heavy-duty vehicles constitute as a major reason that urban NO_2 exceeds the standard in the Beijing–Tianjin–Hebei region.

Adjusting the transportation mix, diverting freight from road to rail, and using heavy-duty diesel-fueled trucks less in the Beijing–Tianjin–Hebei region are the key to air quality improvement. To reduce road freight and nitric oxide emission and to thus improve air quality, MEP has in recent years, with the vigorous support of transportation and public security departments, worked on the following tasks:

First, stronger environmental regulation on vehicles has been imposed. In April 2017, the Ministry of Public Security issued a notice stating that a nationally unified code for emission exceeding the standard would be added to the traffic violation penalty system. Also, a joint law enforcement mechanism whereby environmental departments collect evidence and public security departments impose punishments has been established and effectively implemented. You may have noticed that since the beginning of this year, Beijing has been conducting law enforcement inspections that focus on regulating the emission of heavy-duty diesel-fueled vehicles. Altogether 14.726 million inspections of motorized vehicles have been made, which spotted 18,000 violations, 2.3 times that of the same period of last year.

Second, rail is replacing road in transporting coal between its destinations or origins and ports in Tianjin, Hebei, Shandong and the Bohai Rim area. Currently, this replacement is already completed in Port of Tianjin, Huanghua, Tangshan, Qinhuangdao, Weifang, Yantai, etc. In addition, Tangshan and Handan in Hebei Province and Binzhou and Liaocheng in Shandong Province have initiated the construction of roads connecting railway with steel enterprises and electrolytic aluminum enterprises that need a large amount of materials delivered to them. According to statistics, such companies that use rail to transport materials can cut the trips of heavy-duty diesel-fueled vehicles by tens of thousands a day.

Next, MEP will push and work with relevant departments and localities to adjust transportation structure as a major move to control air pollution. Specifically speaking, the adjustment means enhancing rail freight capacity, improving rail transportation service, facilitating multi-modal transportation involving containers, shipping, and railway, and increasing the share of railway transportation at a faster pace. It also means encouraging the development of clean cargo fleet, implementing off-peak transportation and banning the use of diesel-fueled trucks during heavy pollution prewarning period. In addition, the adjustment requires us to accelerate the building of an interconnected remote sensing monitoring network with shared monitoring and access and to take 24-hour comprehensive management and control measures against high-emission vehicles including diesel-fueled ones: once vehicles violate laws or regulations in a place, for example, excessively emitting or overloading, they will be subjected to nationwide punishment.

China Business Network Daily (CBN Daily): At the beginning of this year, the CPC Central Committee and the State Council issued the notification of establishing Xiong'an New Area in Hebei Province, which is hailed as a thousand-year project and national strategy. What has MEP been doing to facilitate the building of Xiong'an New Area?

You Yanxin: Let me answer your question. The CPC Central Committee and the State Council decided to establish Xiong'an New Area and highly values its ecological and environmental protection. In the meeting on the Planning and Development of Xiongan New Area held at this February, President Xi Jinping stressed seven key tasks, one of which is to "develop Xiong'an into a beautiful, bright and clean ecological city with blue skies, green trees and clear water." The 19th CPC National Congress report proposed positioning Xiong'an New Area as the key that takes over the non-essential functions of Beijing as the capital, promoting the coordinated development of the Beijing–Tianjin–Hebei region, developing forward-looking plans and adopting high standards for building the Xiongan New Area.

MEP has been earnestly implementing the arrangements of the CPC Central Committee and the State Council and actively cooperating with and assisting Hebei Province and Xiong'an New Area in relevant matters, mainly the following three:

First, MEP has strengthened leadership and set up a work mechanism. MEP held a Leading Party Members' Group meeting to discuss the ecological and environmental protection of Xiong'an New Area and established a leading group for that, the head of which is Minister Li Ganjie and deputy heads other ministerial leaders. Also, MEP has issued the *Implementation Plan for Promoting the Ecological and Environmental Protection in Xiong'an New Area in the Near Term*, specifying the tasks and timetables. As I just told you, on November 13 MEP and the people's government of Hebei Province signed a strategic cooperation agreement on the ecological and environmental protection in Xiong'an. MEP is now assembling the Xiong'an ecological and environmental protection technical advisory committee.

Second, MEP has been vigorously solving major problems. Central authorities have earmarked 500 million yuan from the central water pollution prevention and control special fund to support comprehensive environmental remediation in Xiong'an New Area. MEP is the first to thoroughly investigate soil pollution in Xiong'an, and the investigation of construction land for such places as hospitals and schools was prioritized and has been completed.

In addition, since this May, MEP has, in accordance with the instructions and requirements of the State Council, worked with the people's government of Hebei Province to transfer competent personnel to carry out special environmental inspection of Xiong'an New Area. The inspection did away with altogether 12,098 enterprises that are in unauthorized locations, lack the proper certificates and fail to meet emissions standards and handed over 674 cases to local authorities for rectification, 636 of which have been completed and 38 are ongoing. It is fair to say that the inspection has cracked down hard on all kinds of environmental illegalities in the new area and thus contributed to its environmental improvement.

Third, MEP strives to do a good job in systemic planning, top-level design, instructions, and technical support. Currently, there are two plans on the ecological and environmental protection of Xiong'an New Area. The first is the *Xiong'an New Area Ecological and Environmental Protection Plan*, the formulation of which is led by the Chinese Academy for Environmental Planning of MEP, entrusted by the Environmental Protection Department of Hebei Province. The second is *Baiyangdian Ecological and Environmental Control and Protection Plan*, which the Research Center for Eco-Environmental Sciences under Chinese Academy of Sciences, entrusted by the Hebei Development and Reform Commission and assisted by the Chinese Academy for Environmental Planning of MEP, took the lead in formulating.

Next, MEP will coordinate Hebei Province and the plans' formulating institutions to offer more guidance on the two plans, stress the requirement of prioritizing the delineation of ecological red line and strictly sticking to it and of strengthening ecological space management and control, and place areas with important ecological functions such as Baiyangdian under mandatory and vigorous protection. The aim is to delineate the red line in such a way that it encircles all the important ecological spaces and thus becomes the crucial foundation for Xiong'an to develop itself into an ecological city.

Now Mr. Wang Jinnan, President of the Chinese Academy for Environmental Planning under MEP and chief planner of the two plans, will elaborate on them.

Wang Jinnan: Good morning, friends from the media. I would like to add something about the progress of the *Xiong'an New Area Ecological and Environmental Protection Plan*. The Leading Party Members' Group of MEP highly values its formulation. Upon taking on the task, MEP assembled a competent team including members from the Chinese Academy of Sciences, Peking University, and Tsinghua University and sent it to Xiong'an. The team has carried out survey, research, and planning, then accordingly produced an initial plan, and is aligning it with the overarching plan.

Wang Jinnan, President of the Chinese Academy of Environmental Planning of MEP

Overall, this plan is about fleshing out the CPC Central Committee's general requirements on the planning and construction of the Xiong'an New area, and in particular, the instructions of President Xi Jinping. At present, the plan has been basically completed. Experts are having discussion and debates about it and strive to comprehensively align it with the overarching plan. Regarding the plan, the most fundamental starting point is committing it to "global vision, international standards, Chinese characteristics and high-level positioning," the requirements set for Xiong'an New Area; the goal is to develop the area into a world-class green, ecological and livable city; the basic principles are prioritizing ecology and pursuing green development; resources, ecology and environment and the negative list will be leveraged to optimize the development of the whole new area. We will protect the ecological environment of Baiyangdian and, on the basis of it, optimize the layout of Xiong'an. Also, we will improve the environment of Xiong'an and, capitalizing on this, improve the environment of Baiyangdian and even the Beijing–Tianjin–Hebei region at large. Green infrastructure and intelligent and environmentally friendly services will be deployed to make Xiong'an greener. That means we want to adhere to green development during Xiong'an's construction and operation.

In a nutshell, by preserving ecological space, engaging in systemic protection and restoration, improving the environment, prioritizing green development, pursuing intelligent and environmentally friendly services, and innovating systems and mechanisms, we will build Xiong'an New Area into a green, low-carbon, intelligent, livable and business-friendly modern area with international influence and competitiveness and enabling the harmonious coexistence between man and nature. That is our basic planning. Thank you.

Jiemian News: The report of the 19th CPC National Congress pointed out that by 2035, "the goal of building Beautiful China will be basically achieved." Can China's environmental problems be effectively addressed by 2035 when people can enjoy a good ecological environment with blue sky, green land, and clear water? What are MEP's plans and measures to achieve this vision?

You Yanxin: Thank you for your question. The report of the 19th CPC National Congress puts forward Socialism with Chinese Characteristics for a New Era and specifies the goals of building a moderately prosperous society in all aspects and, on the basis of that, basically achieving socialist modernization by 2035 and developing China into a great modern socialist country by the middle of the twenty-first century.

It is fair to say that the report of the 19th CPC National Congress draws the grand blueprint for building an ecological civilization in the new era and the strategic path to build Beautiful China. By 2035, the ecological environment of our country will be fundamentally improved, and the goal of building Beautiful China be basically achieved. By the middle of this century, China will be developed into a great modern socialist country that is prosperous and strong, democratic, culturally advanced, harmonious and beautiful, and new heights will be reached in every dimension of ecological advancement.

The Leading Party Members' Group of MEP attaches great importance to the new goals, plans and requirements for ecological civilization building and environmental protection stated in the 19th CPC National Congress report. After the release of the report, Minister Li Ganjie immediately instructed relevant departments to deepen research and specify goals, targets and measures. In the subsequent group study activities, he further proposed requirements for implementation.

We are organizing a study group to formulate the eco-environmental protection strategy and implementation roadmap for building Beautiful China in accordance with the spirit of the 19th CPC National Congress. With the goal of building a moderately prosperous society in all respects by 2020 in mind, we will take stock of the major logic, landscapes and trends on China's economic, social, and environmental development as well as global sustainable development. On the basis of that, we will take advantage of China's institutional and political advantages attributable to its socialism with Chinese characteristics to address our ecological and environmental problems. Meanwhile, we will come up with the overarching strategy and framework for eco-environmental protection in the new era, draw a roadmap for realizing eco-environment goals aligned with modernization, identify the problems and approaches, and provide the macro strategic decision-making support for achieving China's socialist modernization in the new era and "Chinese approaches" for global sustainable development.

Specifically speaking, in accordance with the plans and requirements stated in the 19th CPC National Congress report, we focus our studies on several key areas. For example, we have been studying the overall strategy, targets and indicators of building ecological civilization for Beautiful China and strive to develop a detailed target and indicator plan as soon as possible for "fundamentally improving the eco-environment and basically achieving Beautiful China" by 2035. Meanwhile, in light of the four major tasks put forward at the 19th CPC National Congress, we are studying key matters such as the national green development and its implementation path; roadmaps for improving the environmental, protecting and restoring ecology and protecting rural environment as part of the rural revitalization initiative; modernization of the ecological environment governance system and capacity.

Besides, 2035 will be the end of China's 16th Five-Year Plan. We will carry out ecological environmental protection strategy planning and research for the "14th Five-Year Plan" and "15th Five-Year Plan" to lay a solid foundation for the 16th. We would appreciate your attention and support.

China Youth Daily: China is vigorously promoting public–private partnership (PPP), and environmental protection is one of the key areas that PPP is applicable to. What measures has MEP taken in promoting PPP? What are the next steps?

You Yanxin: I will answer this question. Environmental protection concerns public interest, and PPP application to environmental protection does draw much attention. PPP is important to address the investment shortage in environment protection and use funding more efficiently, which in turn contributes to better and more efficient supply of environmental public goods and services. MEP takes PPP seriously and has done a great deal with relevant departments in such aspects as policy formulation, project selection, demonstration, publicity, and guidance. The main tasks are the following four:

First, together with the Ministry of Finance, MEP issued the *Implementation Opinions on Advancing the Cooperation between the Government and the Private Capital in Water Pollution Prevention and Control*. Water pollution can be the entry point for PPP application. The special fund for water pollution will treat water PPP projects more favorably.

Second, together with relevant departments, MEP issued the *Plan of Cultivating and Developing Market Actors for Controlling All Types of Agricultural Pollution from NonPoint Sources and for Handling Rural Sewage and Garbage*. This is one of the plans for eco-civilization institutional reform adopted at the third plenary session of the 18th CPC National Congress. Regions where conditions permit are encouraged to implement PPP projects for county-wide or regionally integrated agricultural and rural environmental control.

Third, we cooperated with relevant departments in issuing the *Notice on Comprehensively Implementing the PPP Model in Sewage and Waste Treatment Projects with Government Involvement* to further encourage the adoption of PPP.

Fourth, the Ministry of Finance and MEP vigorously organize promotion activities for environmental PPP projects, and a large number of such projects have been implemented. We are working with the Ministry of Finance to promote, review and approve the fourth batch of PPP projects.

Currently nationwide, PPP projects in environmental management are vigorously supported by localities. But overall they are still at a very early stage, and there are some difficulties and problems during declaration, design, and implementation, which mainly fall into the following categories:

First, many local governments view PPP as a means of replacing traditional financing platforms and getting national policy support without pursuing environmental benefits as the ultimate goal.

Second, the lack of effective mechanisms tying environmental service price to pollution control performance during implementation makes it difficult to properly assess and constrain private capital.

Third, environmental PPP projects have such common problems as faulty return mechanisms, unclear delineation of responsibilities and slow progress.

To advance the environmental application of PPP, we will target at problems, be goal-orientated, and cooperate with relevant departments to focus on the following three tasks:

First, bolstering policy design. With the prevention and control of air, water and soil pollution in mind, we will do more research on PPP mechanisms, improve the return mechanism, boost assessment and regulation and put in place incentives, restraints and requirements.

Second, allocating more environmental protection special funding to PPP projects. Among projects listed in the national project reserve that MEP and the Ministry of Finance are jointly building and running, PPP will be prioritized.

Third, working on project demonstration and promotion. We will timely draw on previous experience, intensify publicity and bring into play the catalytic role of demonstration projects by selecting the better-performing projects for nationwide promotion. We would appreciate your support. Thank you.

People's Daily: MEP is vigorously promoting permit trading. Some people say that it only has a limited effect. What do you think of it? What experience has been learned from the current emission trading pilots? What are your next steps?

You Yanxin: The permit trading mechanism is an important part of the ecological civilization institutional reform. We launched the scheme ten years ago. Now 28 provinces have pilot projects of payment-based emission and permit trading, 11 of which and Qingdao city have national approval while the rest initiated the pilots by themselves. As things stand, these pilot projects have overall secured initial success on several fronts.

First, the institution has basically been established. The General Office of the State Council issued several guiding documents, one being the *Guiding Opinions on Further Advancing Pilots for Payment-Based Emission and Permit Trading*. Moving forward steadily in accordance with national requirements and local practices, the pilot provinces have established a provincial-level system of payment-based emission and permit trading, and have done a great deal in developing institutions and platforms, addressing technical challenges and innovating in policies. So far an orderly permit trading market has basically taken shape. According to data we collected nationwide, as of August this year, the 11 pilot provinces and Qingdao approved by the state had collected a total of 7.3 billion yuan in emission payment and 6.2 billion yuan from permit trading. The figures in provinces that initiated the pilots by themselves are much lower, only 360 million yuan and 500 million yuan from emission payment and permit trading respectively.

Second, the permit trading pilots have effectively reduced emission. Some pilot provinces have tied the payment-based emission and permit trading to pollution reduction, emission reduction targets for enterprises and environmental protection infrastructure construction. They have put in place a pollution source management system relying on mostly administrative regulation supplemented by market approaches and fostered public awareness that resources and the environment have a price. All these efforts have played an active role in adjusting the structure of high-pollution and high-energy-consumption industries, restructuring the economy, boosting enterprises' ability in controlling pollution and reducing emissions. The quality of the environment has thus been bettered.

Third, the pilots facilitate the innovation in other relevant institutions and policies and thus contribute to environmental management. Local authorities have taken the initiative in permit trading. For example, provinces like Zhejiang and Shanxi have initially established an investment and financing mechanism for enterprises to use their permit as collateral to get mortgages and loans; Chongqing Municipality has, drawing upon the rating of financial assets such as securities, put in place a permit registration mechanism; provinces like Zhejiang came up with the card-swiping pollution emission system.[5] All these efforts have contributed to better environmental management.

Judging from how the pilots have been running overall, payment-based emission and permit trading are still troubled by some tough problems. For example, the current laws and regulations have not provided enough support; the preparations needed for emission permit approval and pricing are not ready; the secondary market for permit trading is not active enough. The majority of transactions now take place in the government-dominated primary market. On top of that, the monitoring and regulation system is yet to be improved. According to the documents released by the State Council, the three-year pilot scheme is coming to an end this year. We are discussing with the Ministry of Finance about the policies for the next stage of the scheme. Facing all these problems, we need to dig deeper, learn from the pilots and concentrate on the bottlenecks affecting some policies.

Guangming Daily: I have two questions. First, in 2016 the Ministry of Finance, MEP, National Development and Reform Commission and the Ministry of Water Resources released *Instructions on Accelerating the Building of the Basin Upstream-Downstream Horizontal Eco-Protection Compensation Mechanism*. So how is the compensation working now? What are the next steps?

Second, the *Ecological Environment Protection Plan of the Yangtze River Economic Belt* has been released and translated the principle of "well coordinated environmental conservation and avoiding excessive development" into a

[5]An enterprise has to swipe a card before pollution emission so that the amount of pollution emitted can be recorded and added accumulatively before the upper limit is reached.

construction roadmap. In implementing the plan, how should localities grasp its requirements?

You Yanxin: Thank you for your questions. The CPC Central Committee and the State Council highly value the compensation between the upstream and downstream basin and have made decisions and arrangements on it in the overarching ecological civilization institutional reform.

Two documents, the *Opinion of the General Office of the State Council on Improving the Eco-Protection Compensation Mechanism* and *Instructions on Accelerating the Building of the Basin Upstream-Downstream Horizontal Eco-Protection Compensation Mechanism* issued by such departments as the Ministry of Finance, National Development and Reform Commission and the Ministry of Water Resources, proposed establishing the upstream-downstream horizontal eco-compensation mechanism and pilots. To that end, MEP and the Ministry of Finance have been working hard in accordance with the decisions and arrangements made by the CPC Central Committee and State Council and have mainly carried out the following three tasks:

First, improving the interprovincial eco-compensation mechanism. MEP and the Ministry of Finance are facilitating the pilots for five cross-provincial rivers: Xin'an River between Zhejiang and Anhui Province, Jiuzhou River between Guangdong and Guangxi, Dongjiang River between Jiangxi and Guangdong, Tingjiang–Hanjiang River and Luanhe River. Work on the pilots has been going on smoothly. The people's governments of provinces and the municipality involved have signed eco-compensation agreements. In the principle of "sharing cost and benefits and cooperating in environmental management" and on the basis of the water quality of cross sections at boundaries, they have been vigorously protecting and managing the basin ecology and environment, effectively improving water quality.

Second, implementing the rewarding mechanism. The Ministry of Finance values and supports this mechanism, and together we have formulated the basic rules for the reward. As long as pilot provinces sign the agreement, they would be eligible for fixed-amount subsidy from state revenue. Assessment and calculations will be made at the end of a year based on water quality monitoring results including that of every monitoring site at the cross section at the boundary. Provinces meeting the assessment target will be rewarded and those failing to do so will face deductions in their compensation fund.

Since 2012, 4.999 billion yuan from state revenue has been allocated as ecological compensation and reward fund. The original amount allocated from state revenue was more than 4.999 billion yuan. The decimal point exists because some money was deducted as penalty to the two rivers that failed to meet environmental targets, one being Tingjiang–Hanjiang River in Fujian Province and the other Jiuzhou River in Guangxi Province. MEP has stepped up its guidance and supervision for eco-compensation and thus supervised and inspected the eco-compensation fund of Xin'an River.

Third, expanding the upstream-downstream horizontal compensation range. To implement the *Ecological Environmental Protection Plan of the Yangtze River Economic Belt*, MEP has cooperated with the Ministry of Finance to push forward this horizontal eco-compensation scheme in the Yangtze River basin. Now the two ministries are studying the fiscal incentive policies for expanding that scheme, an initial plan for which has taken shape. At the same time, we are coordinating Beijing Municipality and Hebei Province to put in place an eco-compensation mechanism for the upper reaches of the Miyun Reservoir. The progress is smooth and both sides are working on it with great enthusiasm. The enormous coordination behind each program launched speaks about the proactive attitude of the upstream and downstream provinces in protecting the environment.

Overall, the basin eco-compensation work is still in its infancy, and we have encountered some difficulties and problems in practice. First, legislation on eco-compensation is still lagging behind, and the standard system is yet to be improved. Second, there can be many means of eco-compensation, but we do not have many at our disposal. Third, coordinating the upstream and downstream and reconciling their interests is still very difficult.

Next, we will mainly do three things. First, we will keep pushing the eco-compensation initiative by expanding pilots in the Yangtze River basin and the Beijing–Tianjin–Hebei region, as I mentioned earlier. We are also studying how to build such an eco-compensation scheme in the Chishui River basin by leveraging the reform of cross-basin law enforcement agencies in that area.

Second, we will do a good job in cross-section water quality monitoring and performance evaluation. Unified standards and regulations will be applied to monitoring and evaluation, and the performance evaluation mechanism will be improved by tying central fund allocation to performance results.

Third, we will improve the supporting system and standard system. MEP will factor in the service value of the ecosystem, the input needed for ecological protection and environmental management, the opportunity cost and the level of economic development. On the basis of that, efforts will be made to shed more light on compensation standards and technical specifications and improve the calculation methods of compensation funds, which will be the technical support that makes the compensation more scientific and fairer.

Wang Jinnan: Let me add something. I think the upstream-downstream eco-compensation mechanism is a major innovation in China. Moreover, this mechanism has been recognized by some international organizations. They hope to draw upon the model and factor in their national conditions as well as the water ecological protection plans for their own basins to establish an incentive mechanism for water protection. So I think the horizontal eco-compensation is a model with Chinese characteristics and we need more bold innovations in that vein.

You Yanxin: The journalist's second question concerns the policy of "well coordinated environmental conservation and avoiding excessive development" for the Yangtze River Economic Belt. As we all know, at the symposium on promoting the development of the Yangtze River Economic Belt held in Chongqing at the beginning of last year, General Secretary Xi Jinping clearly pointed out that we must take a long-term perspective and focus on the interests of the Chinese nation when considering developing the Yangtze River Economic Belt, for which we should prioritize ecology and pursue green development. This set the overall tone for the eco-environmental protection of the economic belt, put us all on the same page and established the top-level design and strategic direction. The *Ecological Environmental Protection Plan of the Yangtze River Economic Belt* manifests the important principles conveyed in the speeches of General Secretary Xi Jinping and the New Ideas, New Thoughts and New Strategies on Governance, and is of major and far-reaching significance to the ecological environment protection and socioeconomic development of the Yangtze River Economic Belt.

In a nutshell, the policy of "well coordinated environmental conservation and avoiding excessive development" boils down to "three waters", "four focuses", and "five features". "Three waters" means working on water resources, water ecology, and water environment, an approach stressed in the *Ecological Environmental Protection Plan of the Yangtze River Economic Belt* to achieve harmony between human and water. Via setting and sticking to the limit of water resources utilization, the red line of ecological protection and the bottom line of environmental quality, we facilitate efforts to meet relevant targets and fulfill relevant tasks to effectively protect and improve water quality.

"Four focuses" refers to the upper, middle and lower reaches, key areas, major engineering projects, and major institutional innovations. The *Ecological Environmental Protection Plan of the Yangtze River Economic Belt* manifests the principle that "mountains, rivers, forests, farmlands, lakes, and grasslands are a life community." The plan stresses the overall protection, systemic restoration and comprehensive management of the upper, middle and lower reaches of the basin; it manages, protects and restores key areas, particularly Dongting Lake, Poyang Lake, and the Yangtze River estuary; geared toward the goal of improving the Yangtze River ecology and environment, the plan devises a number of major projects of strategic importance and integrates major tasks and projects; the plan also introduces reforms and innovations in environmental management, for example, differentiated environmental access, eco-compensation, and innovation in environmental governance system, to produce synergy conducive to environmental protection.

"Five features" means making the Yangtze River harmonious, healthy, clean, beautiful and safe. Based on the targets and indicators set in the *Ecological Environmental Protection Plan of the Yangtze River Economic Belt*, the special conditions of the Yangtze Economic Belt, target feasibility and technical availability, and in line with the "five features" overall framework, a target and indicator system has been established. Specifically speaking, to strive for a harmonious Yangtze River, we need to promote the rational use of water resources and develop a harmonious relationship between the river and lakes; for a healthy Yangtze River, we need to gradually improve ecological service functions such as soil and water conservation and biodiversity preservation; for a clean Yangtze river, we need to continuously improve water quality; for a beautiful Yangtze River, we need to build air, soil and other environmental safety shields; for a safe Yangtze River, we need to ensure the effective control of environmental risks. We hope that through our efforts on "three waters", "four focuses", and "five features", we can truly make the Yangtze River a better place as required by General Secretary Xi Jinping. Thank you.

China News: Recently it is reported that an MEP document specifies that in cases of monitoring data falsification, the highest figure measured in that month will be adopted as the figure for the day the falsification occurs, while previously the highest figure was adopted as the average of that month. This is interpreted as a more lenient adjustment to the punishment of data fraud and causes controversies in the industry. Is there really such an adjustment? What do you think of it?

Liu Youbin: I do not think the conclusion you draw is right. MEP has never shown any leniency toward falsification. We all care about data integrity as it is the lifeline of environmental monitoring. MEP highly values it. After Minister Li Ganjie took office, the first visit he paid was to the National Environmental Monitoring Center where he told all the staff there to ensure that the monitoring data would be "true, accurate and thorough."

In recent years, MEP has earnestly implemented the law, regulations, and requirements such as the *Environmental Protection Law* and *Interpretation Concerning Some Issues on the Application of the Law to Handling Criminal Cases of Environmental Pollution*, the latter being jointly issued by the Supreme People's Court and the Supreme People's Procuratorate. MEP has stepped up data auditing and shown no leniency toward environmental monitoring data fraud, investigating and penalizing every case spotted. We have three trammels at our disposal for those who commit fraud. The first is administrative penalties. In accordance with relevant laws and regulations and depending on the severity of the case, a sanction of demerit recording, major demerit recording, demotion, removal from office, expulsion or administrative detention will be meted out to personnel held accountable; enterprises will be ordered to rectify, limit production, suspend production for remediation or pay a fine, or even be shut down in cases of gross violations. The second is criminal punishment. According to *Interpretation Concerning Some Issues on the Application of the Law to Handling Criminal Cases of Environmental Pollution* jointly issued by the Supreme People's Court and the Supreme People's Procuratorate, falsification of automatic monitoring data by key emitters shall be investigated and penalized as "environmental pollution crime," and if the falsification also constitutes as the "crime of sabotaging computer information system," the emitters shall be penalized more harshly. The third is civil liability. According to the *Environmental Protection Law*, those responsible for the environmental pollution and ecological damage as a result of their fraud shall bear joint liability with others responsible for the pollution and damage, in addition to punishment pursuant to relevant laws and regulations. I believe at the thought of these three trammels, anyone seized with the impulse to commit fraud will think twice and fully consider the consequences that may follow.

Data fraud does not automatically happen, people make it happen. We crack down on the fraud and hold personnel involved accountable because we want to ensure the integrity of monitoring data. The Party and the government highly value the laws, regulations and capacity building in environmental monitoring. Now we have enough means to punish fraud and to hold the counterfeiters accountable. We also have a relatively sound quality control system, which allows us to timely detect, quickly investigate and finally handle data anomalies. On the one hand, we have zero tolerance for fraudulent behaviors. Every environmental staff member should protect monitoring data from fraud in the way they protect their eyes from harm. This is being highly responsible for data integrity and, by extension, for our people. On the other hand, it is necessary to scientifically and accurately assess the extent and scope of the impact of fraudulent behavior on data. Then we should do our utmost to adopt technical means to restore the data as it is in the first place, neither playing up nor playing down the fraud impact so as to objectively and accurately reflect changes in environmental quality.

As we all know, in this September, the General Offices of the CPC Central Committee and the State Council issued the *Opinions on Deepening Environmental Monitoring Reform and Improving the Quality of Environmental Monitoring Data.* In line with the requirements of the document, MEP will investigate and penalize every locality, monitoring agency, enterprise or individual involved in monitoring data fraud that has been spotted, and adopt measures to respond according to the law such as open admonitory talks, administrative accountability and criminal penalties. The investigation results and penalties will be publicized. Should the fraud constitute a crime, we shall hand over the case to judicial organs according to the law, who will hold the counterfeiters criminally accountable. Our purpose is to ensure that environmental monitoring data is thorough, accurate, objective and true.

Southern Metropolis Daily: In this August, MEP issued the *Implementation Opinions on Promoting the Third-Party Control of Environmental Pollution.* What are the concrete promotion measures in store?

You Yanxin: This document is very important to environmental protection. Environmental pollution control is very technical and requires expertise. As social division of labor is getting finer, we should trust professionals to do work in their own profession. It is against this background that the third-party model of environmental pollution control was conceived and it has been moving forward with great momentum in recent years. That is of great significance for our environmental protection work at the current stage for the following reasons.

First, the model boosts corporate pollution control efficiency. Under the third-party model, professional and technical personnel and experienced environmental service companies will take over pollution control matters, reducing enterprises' cost and increasing efficiency.

Second, the model is conducive to the healthy development of the environmental protection industry. The environmental protection service companies have expanded from simple equipment manufacturing and engineering construction to operation management. This has provided new impetus for their own growth, which in turn enables the rapid development of the entire industry and contributes to environmental improvement.

Third, the model is good for improving environmental quality. Pollution-emitting enterprises' responsibility for pollution control, via contract, is transferred to and concentrated in environmental service companies. As the main actors of pollution control, emitters will work with environmental protection departments to supervise the service providers, which can effectively reduce the risks of environmental pollution incidents and thus improve the environment.

The third plenary session of the 18th CPC Central Committee clearly proposed adopting the third-party pollution control model. In 2014, the General Office of the State Council issued the *Opinions on Implementing Third-Party Control of Environmental Pollution*, which made some specific requirements on the matter. In order to implement the requirements of the State Council, facilitate the fulfillment of relevant tasks and especially to guide local governments in this regard, in this August MEP issued the *Implementation Opinions on Promoting the Third-Party Control of Environmental Pollution*. Now third-party enterprises are keen on being involved. However, in some localities, policies and regulations still lag behind and the systems and mechanisms needed are yet to be improved, particularly the safeguard mechanisms. So some pollution-emitting enterprises claim that the outcomes are not always satisfactory. We do have some problems, for example, unclear delineation of responsibilities, lack of industrial standards, information asymmetry and the defective pricing and regulation mechanism.

Market vitality does not seem quite strong. Considering that, we will focus on the following four tasks:

First, guiding and standardizing third-party pollution control. We will keep track of and study precedents for third-party liability determination, guide and standardize the operation of third-party facilities and offer more guidance via supporting policies. We will explore the introduction of third-party payment mechanisms, payment tied to environmental performance and compulsory insurance against environmental pollution to safeguard the rights and interests of both pollution-emitting and controlling entities.

Second, strengthening the regulation of third-party pollution control. MEP will fully implement the relevant requirements of environmental protection laws and regulations, further strengthen environmental regulation and law enforcement, increase media coverage of major typical cases and explore the introduction of supporting policies for regulating third-party control enterprises.

Third, encouraging innovation in third-party control mechanisms and models. MEP will establish a reserve for third-party control pilot projects, compile and publish some relevant cases and guide and encourage enterprises to pursue mechanism and model innovation.

Fourth, encouraging information disclosure, which is the key to third-party control. MEP will accelerate the information disclosure, build a disclosure platform and encourage third-party organizations to actively disclose relevant information on these platforms and subject themselves to the supervision of the public.

Reuters: According to data published on $PM_{2.5}$ for October, only 4 of the "2+26" cities in the Beijing–Tianjin–Hebei region and its periphery met the standard. What is the main reason for that? What specific measures has the government taken? What has been achieved? Will the efforts be stepped up?

Liu Youbin: Thank you for your question. Driving to work this morning, I learned that Beijing Municipality publicized the $PM_{2.5}$ concentration data from January to October this year. As of now, the average is 60 $\mu g/m^3$, the lowest compared with the same periods in history. As the weather conditions in Beijing were unfavorable and the $PM_{2.5}$ concentration was very high in January and February, we had to work very hard in months that followed to get this average concentration.

In short, despite the unfavorable circumstances in January and February when the $PM_{2.5}$ concentration rose by 23.5% year on year, the "2+26" cities took great pains and turned the situation around. From March to November 15, the $PM_{2.5}$ concentration fell by 9.8% year on year. Since the stepped-up action for autumn and winter air pollution control kicked off on October 1, the $PM_{2.5}$ concentration in these cities dropped to 64 $\mu g/m^3$, down by 15.8% year on year.

These data are enough to show the effectiveness of our response. Recently people often post pictures of blue skies on their social media accounts. They have intuitively felt that we have more days with blue sky this winter than last and the impact of heavy pollution is subsiding. As we all know, we have just experienced a heavy pollution in the Beijing–Tianjin–Hebei region and its periphery from November 4 to 7. When we released heavy pollution prewarnings, many people could not see the point because the sky was blue at that time. This is because our monitoring, forecasting, and prewarning capacities give us the confidence that our prewarnings can get people prepared and tell localities to take measures in advance to reduce the peak value of pollutants. Monitoring shows that the $PM_{2.5}$ concentration growth is slower this time, weakening the momentum of pollutant accumulation. Experts have analyzed this and conclude that emission reduction measures taken in response to heavy air pollution emergencies are working very well, as evidenced by the 20–30% reduction in the emission of major pollutants.

Therefore, the changes in the monitoring data, our response to this heavy pollution process and the genuine feelings inspired in people by the better environment have boosted our confidence and resolve in preventing and control air pollution in autumn and winter and addressing the critical challenge; these factors also corroborate our faith that the measures we have formulated for autumn and winter are effective and targeted. As long as we continue to tread this path, I believe the regional air quality will be improved.

Of course, localities differ in benchmark data, degree of dedication and pressure conveyed by upper level authorities. Thus, in some cities, air quality does not improve much and even fluctuates. But overall, the regional ambient air quality has indeed been changing for the better since the stepped-up action was launched.

If some localities are not doing a good job and do not fully implement the stepped-up action, and they thus fail to improve air quality, become laggards in the region and affect people's sense of happiness, we will impose quantified accountability on them. There is a saying in China that "there is no jesting in war," We will strictly follow the quantitative accountability plan, investigate and hold accountable the leaders of underperforming localities and impose the pressure on them, ensuring that all the measures of the stepped-up action are well implemented.

The Economic Observer: Environment management has been continuously strengthened in recent years. As investment in environmental protection is an important foundation to it, could you please brief us about such investment in recent years?

You Yanxin: Thank you for your question. As you said, the CPC Central Committee and the State Council have attached great importance to environmental protection over these years, and investment in environmental protection has been increasing. We can see mainly three trends:

First, the investment sum is increasing. During the 12th Five-Year Plan period, the total investment in environmental protection reached 4.17 trillion yuan, up by 92.8% over that in the 11th Five-Year Plan period. The annual growth rate during this period was approximately 10%. The investment made in 2016 was about 922 billion yuan, up by 4.7% year on year. During the 12th Five-Year Plan period, public investment in environmental protection amounted to about 839 billion yuan, the average annual growth rate being 14.5%. Of the 839 billion yuan, 180 billion was earmarked from state revenue as special funding for environmental protection, up by 140% over that in the 11th Five-Year Plan period. State revenue allocated to environmental protection has been rising annually, especially in 2016 and 2017 following the release of the *Action Plan on Air*, the *Action Plan on Water* and the *Action Plan on Soil*, and it is expected to reach 50 billion yuan this year.

Second, the investment structure is being optimized. To support key projects listed in action plans on air, water and soil pollution, the central government has concentrated funds to back projects related to air, water, soil, rural areas, mountains, rivers, forests, farmlands, and lakes. Special funds were set up for water, air and soil pollution prevention and control, rural environment improvement, and key projects of ecological protection and restoration, respectively. As for optimizing funding allocation, we tried to reduce interference from human factors mainly through factor analysis and competition-oriented approach to project approval. In addition, the *Implementation Rules on Incentives of Environmental Protection Special Funds* was released, based on which rewards were granted to areas that made fast progress in environmental management projects and those that saw air improvement in key zones or water environment in key basins in the previous year. This process helps us optimize relevant management mechanisms.

Third, the market-oriented reform is deepened. As I just said, MEP has recently released opinions on putting in place the third-party pollution control. It has also, together with the Ministry of Finance, aired opinions on advancing PPP, offered guidance to local governments and helped build mechanisms that can attract private capital to the environmental protection market.

The abovementioned measures ensure that our environmental quality improvement work is well funded, and indeed good results have been seen in environment quality. In 2016, the $PM_{2.5}$ concentration in the Beijing–Tianjin–Hebei region, Yangtze River Delta and Pearl River Delta dropped by 33%, 31.3% and 31.9% respectively over that in 2013; of the 29 provinces backed by water pollution special funds, 20 have met or even exceeded their water quality target of the year; in the 85 prefecture-level cities backed by special funds, the water quality class of 172 water cross sections has been elevated. In 2016, to implement the *Action Plan on Soil*, the central government integrated funds to set up a special fund, an important safeguard to improve soil quality and prevent and control risks to soil.

However, we have also realized that several prominent problems exist in allocating and using environmental investments from state revenue. First, the total funding falls short of demand. There are still many unresolved issues, and there is a big gap between the investment already made and the investment needed. In 2016, China's investment in environmental protection accounted for 1.3% of the GDP, which is low by international standards, and direct investment in pollution control facilities only accounted for half of that investment.

Second, investment efficiency is not high. Construction is prioritized over operation, investment in operation and maintenance is insufficient and monitoring results are not satisfactory. The environmental performance of some projects fails to live up to expectation.

Third, the market mechanism of investment in environmental protection is not sound enough: the investment return mechanism and the green financial system are yet to be improved. Difficulties and high cost in financing affect the investment readiness of the private sector.

Currently, environmental pollution is still severe in China and pollution control is one of the Three Critical Battles[6] raised in the report delivered at the 19th CPC National Congress. We still have a long way to go. The environmental status quo is far from what people expect for a beautiful life. China is still in a critical period of strengthening weak areas in environmental protection. To turn around the situation in eco-environmental protection and build Beautiful China, MEP will step up its efforts in the following four aspects of environmental protection investment:

First, placing greater emphasis on the fulfillment of responsibilities and increase of investment in pollution control on the side of enterprises. For that, we will delineate enterprises' legal responsibilities, strengthen environmental regulation and law enforcement, continue to carry out annual activities to implement the latest version of the *Environmental Protection Law*, and strengthen coordination between administrative law enforcement and criminal judicial agencies. Also, we will establish a mechanism for rating the environmental credit of enterprises and listing illegal pollution emitters.

Second, making full use of market-based mechanisms to attract private capital. We will vigorously promote the PPP model and third-party pollution control, and improve the pricing mechanism of wastewater, garbage, etc. We will refine the green finance mechanism, and facilitate innovation in mechanisms of resource and industrial integration and investment return, for example, creating portfolios of government-funded and private funded projects.

Third, increasing investment and concentrating funding to resolve prominent environmental problems in key areas. Currently, smog, a public concern, black and putrid water bodies, drinking water sources and rural environment are weaknesses that we will provide greater support to address. Key areas will all need greater investment, for example, the Beijing–Tianjin–Hebei region, Yangtze River Delta,

[6]The Three Critical Battles which China is determined to fight are preventing and resolving the major risks, conducting targeted poverty reduction and controlling pollution.

Pearl River Delta, the Yangtze River Economic Belt, areas along the South-to-North Water Diversion Project, national-level impoverished areas and ecologically fragile areas.

Fourth, establishing a performance-oriented mechanism to use fund more efficiently, specifically speaking, a mechanism for earmarking, adding and deducting special funds based on results of regular performance evaluation. We have been exploring the matter the whole time. We hope that the mechanism can be well implemented and thus tie performance evaluation results of environmental protection projects to funding arrangement.

National Business Daily: The *Opinions of the CPC Central Committee and the State Council on Accelerating the Development of Ecological Civilization* proposes boosting the market-oriented mechanism, which is an inherent part of promoting environmental control and enhancing control efficiency. What has MEP done to promote this mechanism? What are the next steps?

You Yanxin: I would like to direct the question to Mr. Wang Jinnan, President of the Chinese Academy for Environmental Planning.

Wang Jinnan: Policy approaches to protect the environment are, on one hand, mandatory legal approaches such as regulations and standards, and on the other hand, market approaches like environmental economic policies. MEP highly values the role of the latter.

First, MEP, together with relevant departments such as the Ministry of Finance and the National Development and Reform Commission, has introduced and implemented a series of important environmental economic policies over these years, including environmental protection tax, pollutants emission trading, ecological compensation, and green finance.

Some of the policies are influential worldwide, for example, the environmental protection tax. It is the world's first independent national-level tax concerning environmental protection and is a significant part of ecological civilization institutional reform.

Second, MEP is coordinating with relevant departments to steadily promote the application of policies, for example, fiscal, trading, credit, securities and insurance policies. For example, a catalogue has been established listing products that cause serious pollution and pose serious environmental risks. All these policies play a positive role in advancing structural adjustment.

Third, MEP has made great policy innovations, for example, the aforementioned ecological compensation between the upstream and downstream of river basins, payment-based pollutant discharge permit and the trading of such permits, and the reform of environmental protection tax. Such policies mainly aim at prompting emitters to consider their pollution from the economic perspective and calculate the cost themselves before they decide what approaches make the most sense economically. The question of how to get enterprises better internalize the externalities of their pollution emission and at the same time reduce enterprises' costs drives us to seek answers through institutional innovation.

The market mechanism should play an increasingly important role in future environmental protection. That needs us to continue with our reform and innovation endeavors.

Liu Youbin: Today's press conference has come to an end. Thank you.

Regular Press Conference (December)

December 28, 2017

On the morning of December 28, the MEP held the December press conference. Mr. Liu Changgen, Deputy Director General with the Office of National Environmental Protection Inspection and Mr. Liu Youbin, Counsel (Director-General level) with the Department of Communications and Education briefed on the latest developments respectively on the central environmental protection inspection at stage I, and on the recent work of MEP, and then took questions from the media.

Ministry of Environmental Protection of the People's Republic of China,
2017 Press Conference Records of Ministry of Environmental Protection, the People's Republic of China, https://doi.org/10.1007/978-981-13-7330-5_13

MEP December Press Conference (1)

MEP December Press Conference (2)

Liu Youbin (Moderator): Good morning, friends from the media. Welcome to today's press conference.

Central environmental protection inspection is an important institutional arrangement aimed at advancing ecological progress, and received much public attention. Now the fourth of the Stage I central environmental protection inspection is in process with feedback to be provided to relevant local authorities. The end of fourth round also marks a full coverage of China's 31 provincial-level regions, autonomous regions and municipalities across China. We saw much media coverage on the work of the inspection teams, which showed your care and support for this. There have also been in-depth report from the media on our feedback to the localities, which creates a sound environment for guiding public opinions. We hereby express our gratitude to your efforts.

Today we are very honored to have Mr. Liu Changgen, Deputy Director General of the Office of National Environmental Protection Inspection, with us to respond to questions of your concern.

I'd like to brief you on some of the work priorities of the MEP.

15 provinces have formulated their respective plans on delineating ecological conservation red lines.

This February, the general offices of the CPC Central Committee and the State Council have circulated *Opinions on Delineating and Strictly Enforcing Ecological Conservation Red Lines*, which laid out general arrangements on it. The MEP worked together with the NDRC and other departments to guide the local efforts in accordance with the documents. Localities attached high importance to it, setting up their respective leading team led by heads of the provincial party committee and government. They also organized an expert panel mainly engaged in the specific delineation work.

As of today, 11 provinces across the Beijing–Tianjin–Hebei Region and the Yangtze River Economic Belt, together with other 15 provinces including Ningxia Hui Autonomous Region, have all formulated their respective delineation plans, and have which approved by the people's government at the provincial levels. From Nov. 28 to 30, an interministerial coordination group on ecological conservation red lines deliberated and adopted relevant plans of the 15 provinces, which will be issued and put into action by each locality after approved by the State Council.

Looking toward 2018, we will continue to advance the red lines system in the remaining 16 provinces together with other relevant departments, trying to align all local efforts and bring them under uniform national control. Meanwhile, there will also be steps to demarcate the boundaries of the areas protected under the red lines system, and to study and introduce an *Interim Measures on Management of Ecological Conservation Red Lines* together with relevant departments. All these will serve to ensure by 2020 red lines are delineated across China, a basic ecological conservation red lines system set in place, ecological space optimized and effectively protected, ecological functions kept stable, and systems in upholding national ecological security further refined, to bring forward an even more beautiful China.

Now let's give the floor to Deputy Director General Liu Changgen.

Liu Changgen: Good morning, friends from the media. Thank you all for your long-term care and support for our work on environmental protection, especially on environmental protection inspection.

In August of 2015, the central government circulated the *Plan on Central Environmental Protection Inspection*, and in the same year in December, pilot inspection was first initiated in Hebei Province. With 2-year efforts, inspection achieved a full coverage of the total 31 provinces, municipalities and autonomous regions. Thanks to the strong leadership of the CPC Central Committee and the State Council, active public participation, and the support of relevant departments of various levels, central environmental protection inspection has made remarkable headways. As is pointed out by Minister Li Ganjie, the central environmental protection inspection successfully resolved the issues, therefore receiving more "thumbs-up" from the public, recognition from the central government, and support from localities. I think such notable results, in a final analysis, are by and large attributed to the following.

First, the CPC Central Committee and the State Council gave high priority to ecological civilization and environmental protection, advancing which with strong leadership, science-based plans and great resolution. The central environmental protection inspection system is the one that is proposed and promoted by President Xi Jinping himself. He has provided important instructions in every key step, and reviewed each inspection report. He not only pointed the way for environmental protection inspection and explicitly stated the requirements, but also encouraged us to keep on. Other central leaders like Premier Li Keqiang and Vice Premier Zhang Gaoli also made instructions on multiple occasions to guide out efforts. It's the core that guarantees the environmental protection inspection is fully implemented.

Second, the general public trusted us, actively participated in this process, and strongly supported our work. In carrying out the inspection, we always put people first. Those seemingly tiny issues of people's concern were all paid close attention to in our inspection. By taking concrete steps to address these environmental issues troubling people's daily life, we delivered to public a sense of fulfillment and motivated public participation. These several rounds of inspection further advanced information disclosure and publicities on our inspection, guaranteeing their rights to know and to oversee. By this, we earned public trust, with each round of inspection receiving more public tip-offs on environmental violations than the previous one. It's the participation, recognition, and trust from the public that makes the environmental protection inspection more targeted, in-depth, effective, influential and penetrating.

Mr. Liu Changgen, Deputy Director General with the Office of National Environmental Protection Inspection

Third, the local party committees and governments fully understood and supported us, resolutely taking on their political responsibilities in helping with inspection and rectifying the inspected issues. It's clear to all that inspection is mostly about identifying problems and strengthening accountability, which puts a huge pressure on local party committees and governments. With the inspection for stage I coming to an end, we realized the local party committees and governments did treat this seriously, and actively supported our work. It ensured a successful inspection, and as well showed us their determination in fulfilling political responsibilities. They took serious measures to crack the hardest nuts, committed to fixing the detected problems and holding accountable those relevant parties. All these helped address a large number of specific environmental problems, and as well further cemented the mechanisms by which both party and government officials are held equally responsible for environmental damage and the leading officials must take the responsibility for environmental protection within their jurisdiction. All these helped to attain the expected goals of environmental protection inspection.

Fourth, the leaders and staff members of the inspection teams were loyal to their mission and worked very hard to it. Each inspection team has established their interim party group when in inspection, to learn in-depth the General Secretary Xi Jinping's strategic thinking on ecological civilization, and ensure they take a clear political stance and strengthen political awareness. The chief and deputy chief of each inspection team were required to identify priorities and set right direction for inspection, and strictly regulated the entire inspection team. The staff members of each inspection team followed orders and instructions, and devoted themselves to the inspection work, ensuring the inspection was implemented in a strict, standard and authoritative manner.

Environmental protection inspection is a major institutional setup in moving forward ecological civilization and environmental protection. The party group of the MEP highly valued it, led and organized relevant efforts, refined the inspection mechanism, and earnestly put it into action in line with the thinking and instructions of President Xi Jinping and the decisions and plans of the CPC Central Committee and the State Council. In this effort, we won great support from the CPC Central Committee and relevant state departments, as well as care and participation from friends with the media. Thank you.

Liu Youbin: Now the floor is open.

National Business Daily: Given that the central environmental protection inspection involves a spectrum of departments and localities, how do you coordinate and organize relevant efforts? And how did different departments work together with each other and play a role in the inspection?

Liu Changgen: In February 2016, the State Council set up a Leading Group on Environmental Protection Inspection, which consists of members from the MEP and the other relevant 14 departments under the CPC Central Committee and the State Council and is mainly responsible for crafting inspection plans, coordinating specific inspection work and reviewing the inspection reports. With the inspection work unfolding, we've received guidance, support, and assistance from relevant departments of the CPC Central Committee and the State Council. Without them, the inspection can never be such a success.

The General Offices of the CPC Central Committee and the State Council cared for and guided our inspection, sending personnel to join the inspection team and making clear the specific arrangements, priorities, and procedures of inspection. The Office of Central Leading Group for Comprehensively Deepening Reform and that of the Central Leading Group for Financial and Economic Affairs both took environmental protection inspection as a major step toward ecological civilization, which guided the ongoing inspection work, made the overall coordination and put forward specific requirements. The Organization Department of the CPC Central Committee helped to set in place a system for selection of inspection team leaders, reviewing qualifications of the leader candidates and then submitting to the CPC Central Committee for approval. The feedback of the inspection is one of major criteria in evaluating local authorities' performance. The Publicity Department of the CPC Central Committee increased publicity and report of inspection work, with the CPC mainstream media well reporting it in each important milestone, playing a positive role in guiding public opinions. The State Commission Office for Public Sector Reform (SCOPSR) supported, coordinated and assisted us in setting up the Office of National Environmental Protection Inspection, upgrading the major inspection bodies in localities from a public institution to an administrative organ, and establishing various inspection teams. Meanwhile, the Ministry of Supervision (MOS) helped with the transfer of environmental violators found in inspection to relevant departments and held accountable for those who damaged the ecological environment. Other state departments like the NDRC, the Ministry of Finance (MOF), and the National Audit Office all actively support our work, providing many reference materials and useful clues to environmental problems, which is a strong guarantee for us to drive forward inspection.

The central environmental protection inspection is by no means the MEP's solo, but a result of collective efforts of relevant departments. Such a successful fruit of inspection is indispensable from support, guidance, and assistance from these departments. Thank you.

China National Radio: The accountability system is a distinct feature of central environmental protection inspection. Each time when relevant information is disclosed, there will be great social sensation. Yet there's as well some doubts. Some claimed that words on accountability always speak louder than its actions, for those who are held accountable are usually at the most grassroots level. Others claimed that only the information on the number of people held accountable made public without specific names, triggering many doubts on its authenticity. There's also doubt on the fairness of accountability, saying the accountability more often goes to other departments than the environmental protection authorities themselves. What's your response to the above claims?

Liu Changgen: Fixing problems and strengthening accountability are serving as a pair of wheels spinning forward together, and the twin wings of a flying bird, pushing the environmental protection inspection ahead, thus must be advanced simultaneously. Our inspection aims at achieving dual goals, addressing problems on the one hand, and developing it to a long terms mechanism on the other hand. Without either, the goals can hardly be achieved. Usually, accountability is brought into force in central environmental protection inspection in the following three major situations:

First, localities would hold accountable those who fail their duties in addressing the problems reported by the public. This usually takes place amid the ongoing central inspection, which allows public to file complaints and requires the localities to take immediate steps under the supervision of the inspection team. In stage I, we have held a total of 18,199 people accountable, including 875 director-level officials, 6,386 division-level ones, and 10,938 others based on our further analysis. As these issues are mostly found around communities, those who are held accountable are usually at the community level. Judged from current progress, accountability in this case would not only help to address specific environmental problems, but also enhance environmental protection awareness of officials at the community level, whose inaction, misconduct, or abuse of power are usually considered the last-mile problem. With these, we can clear the way ahead.

Second, when the inspection mission ends in a locality, the inspection team would hand over to it the cases to be dealt with for ecological and environmental damages. For the Stage I, the inspection teams have handed over 387 cases where people are liable for ecological and environmental damages. In such case, those who assume the leadership, management and supervision roles usually bear the liability, which forces the leaders of local party committee and government to take on their political responsibility in promoting ecological civilization. We have made public the results of such accountability in 8 provinces (autonomous regions) covered by the first round of inspection, which involved 130 director general and deputy director general level officials including 24 director generals. To date, localities have successfully held to account those who are involved in cases transferred in the second round, with an even tougher measure than in the first round. We will soon make public the results.

Third, localities take a proactive approach to hold to account those who fail to rectify the problems, especially the ones identified before, ordered for rectification, but remaining unsolved or back to former conditions upon the review of a target inspection team sent by us targeting key problems in key areas. Accountability in such case is intended to further strengthening the local responsibility in rectifying problems to expedite the rectification process. A good example is the problem that inspection team found in Nantong city, which was circulated to the public on December 25. We received public reports on it, and then sent an inspection team there, and handed over the results to relevant departments. Above are the three major conditions under which the accountability take place.

You also mentioned holding to account those other than environmental protection authorities. It's in my view easy to understand, as President Xi once mentioned that we must adopt a holistic view to conserving our mountains, rivers, forests, farmlands, lakes, and grasslands, which together are a community of life. In this connection, central environmental protection inspection is a comprehensive one, going beyond the environmental domain. Therefore, be it an environmental authority or from other departments, once they fail their duties for environmental protection, shall all be called to account. It's also a basic requirement of central environmental protection inspection. Take the first circumstance under which accountability is required as an example where rectification goes side by side with the inspection. Statistics shows that of those held accountable, 1,488 are leaders of local party committee, 4,254 from the government, 6,638 from relevant management departments, and around 6,000 are heads of state-owned enterprises, communities, and village party committee. Of those from relevant management departments, we have 2,612 from department of environmental protection, 923 national land and resource, 904 forestry, 580 water conservancy, 433 housing and urban–rural planning, 369 agriculture, 337 urban management, 159 work safety, 155 industrial and informational technology, 118 transport, 56 public security, 50 development and reform and 42 tourism, basically covering all authorities associated with environmental protection. It's also in line with the requirements that both party committee members and government officials are held equally responsible and leading officials must take the responsibility for environmental protection within their jurisdiction. There's never a bias against any one.

The MEP has a list of officials held accountable, which contains information like the name, title, position, as well as for what reason and in what way one is held to account. But in the end it's the localities' job to disclose it and decide how much would be released. We admit that in the very beginning, we saw very few information disclosed, and that's why we later in the third and fourth rounds of inspection requested the localities to further open up information on typical cases. Thank you.

People's Daily: We saw a lot number of issues found in almost each round of inspection at stage I. Some are very prominent, receiving much public attention after being disclosed, and some of which are quite common in localities. In your view, what are the typical issues often found in localities?

Liu Changgen: In our inspection to various localities, we did find quite a few environmental issues that are typical and share a common nature. The central leaders are very concerned about them, specially requesting us to analyze and sort out these issues. We then come to the following conclusions based on our analysis:

First, we found prominent air and water environmental problems in some localities, where air pollution was yet to be effectively controlled and prevented, and some watersheds especially the tributaries were heavily polluted. It's partly attributed to the poor local industrial structure, energy mix, and development mode, and greatly concerned by the public.

Second, it's still common in localities where environmental infrastructure on pollution control far lagged behind, sewage was directly discharged to rivers and sea, solid waste was poorly managed, black and putrid waters remain unsolved. In the inspected 35 priority cities, the daily discharged domestic sewage amounted to over 12 million tons in an accumulative term. Some told me recently that environmental management in urban areas had made great strides. However, results of our inspection indicated that we are far from there yet, especially when it comes to the infrastructure construction. That's all for the second category of problems common to localities.

Third, some nature reserves were mostly found with problems like approving and constructing development projects against the laws and regulations, mineral mining, quarrying and sand mining damaging regional ecosystems, and developing without follow-up restoration. In some localities, a considerable sum of reserved fund required from the enterprises for restoring and conserving the geological environment of mines were left untouched, while the damaged ecosystems in those old and waste mines remain without any improvement. Clearly, their awareness of environmental protection shall be further enhanced.

Fourth, problems like overuse of water resource, turning lake into other use, damming across a river, encroaching coastline, damaging wetland, and illegal reclamation of land from sea are often found, which yet to be taken seriously by the localities. We saw more issues concerning reclaiming land from sea reported by the fourth round of inspection than the other three rounds, which were also widely reported by the media. These issues are usually spotted in Hainan and Shandong Provinces.

Fifth, industrial pollution remains an acute problem. First, it's mainly seen in industrial parks, where industrial enterprises are mostly concentrated, and so is the pollution. Such problem has already been found in Jingdezhen at Jiangxi Province and chemical industrial park of Guannan County of Lianyungang city at Jiangsu Province, and we have respectively held admonitory talks and circulated the found problems to public. Second, it's also associated with companies in unauthorized locations, lacked the proper certificates and failed to meet emissions standards. They widely spread in both Beijing–Tianjin–Hebei region and other regions and caused heavy pollution, an issue frequently reported by the local people where the inspection team stationed. Therefore, dealing with industrial pollution remains a top priority of overall pollution control efforts.

Sixth, rural environmental pollution stood out as a major problem. Both pollution control and environmental management there are well behind, with solid waste and sewage left untreated everywhere. It's further worsened with lack of attention, supervision, inspection, and solution as well as inaction of local authorities in dealing with rural environmental issues.

A couple of days ago, I came across an online article analyzing 16 of those cases brought up by our inspection, and carefully went through it. It came to the similar conclusion as I mentioned above. In my opinion, these issues are on the one hand due to the fact that China is still at the developing stage right now, while on the other hand the result of the indifference, inaction, and irresponsibility of localities. Thank you.

China Times: We noticed that when the inspection team is there, the localities paid extra attention, working actively to support the inspection and fix the found problems. Yet it also triggered concerns that everything might go back to normal as the inspection team gone, the effect of inspection can be significantly weakened. Could you please share with us the current state of localities in dealing with the found problems?

Liu Changgen: We also share your concerns on this aspect. At the very beginning of the inspection, namely in the first and second rounds, we did see some problems that were once found, rectified, yet in the end came back in a few localities. We also received complaints saying that problems were tackled when the inspection team was there, but surfaced again as soon as the inspection team left.

We came up with a number of measures to address it. We requested localities to submit monthly checklists on the developments in rectifying problems. The following are required to be specified in the checklist, including, first, instructions and feedback of local party committees and governments on rectification work and on-site supervision; second, long-term mechanism brought forward by localities including some important policies, regulations and systems; third, of those dozens and even hundreds of problems left for localities to deal with, the progress in fixing some of the most significant ones, which will be closely followed by us; fourth, updates from some departments and cities that are considered main parties liable for those problems on progress of making rectifications; and fifth, developments on how local party committees and governments exercise supervision. We try to make the checklist as much specific as possible, and will call supervision bureau to verify some of these reports. If a party fails to fully rectify the problems, we will issue warnings, urge for more steps, and take even tougher measures.

We also have instituted targeted inspections particularly for reviewing problems once identified in key regions and key sectors. This is the case in Nantong city which we sent targeted inspection team there, found the problems remain unsolved and thus held relevant officials liable. We also worked out some punishment mechanisms like issuing notice, circulating the found problems to public, holding admonitory talks, and transferring the party liable to other departments, which helped the localities a lot in putting in place and enhancing the long-term mechanism to rectify problems as required.

We have also strengthened information disclosure and publicity. The information to be disclosed by localities must come into great detail, including rectification plan, how problems are fixed and how work is implemented, to ensure an effective public oversight. This whole system is designed to fully play out the role of inspection, which won't be stopped until the problems are rooted out. Our survey also indicates that localities are attaching high importance to resolving problems. They took concrete steps and made solid progress, which are shown as follows.

First, local authorities led the efforts to achieve the goal. Local party committees and governments tasked with rectifying relevant problems all saw it an important political task, a significant project concerning people's well-being, and an issue concerning overall development that needs to be focused on and actively implemented. Provinces have set up leading teams on it, with half of which having heads from both party committees and governments as co-leaders, and the rest having a head either from party committees or governments as leader. For example, Hubei Province took a "delivery-against-payment" approach, by which the municipal official sign a letter of commitment first, promising to the provincial officials to finish the tasks within a specified period of time, and if the results are not delivered in time, the former shall be held accountable. It's a very useful mechanism, translating the huge pressure on municipal officials into concrete and quick steps.

Second, implementation was expedited. To date, of the 1,532 items called for rectification in the 22 provinces inspected by the first three rounds of inspection, 639 items have been completed, registering a completion rate of 42%. So far, environmental problems identified in the Mount Kalamaili Ungulate Nature Reserve in Xinjiang have been basically fixed, the total area of wetland in the Hulun Lake in Inner Mongolia significantly expanded, lakes previously claimed for aquaculture returned in Hubei Province, pollution control in the 9 major plateau lakes in Yunnan Province, Dongting lake of Hunan Province and Poyang Lake of Jiangxi Province speeded up, and encroachment on the wetlands across the Chaohu lake fully banned in Anhui Province.

Third, long-term mechanisms have been further improved. The inspection and the work to rectify inspected problems is also an opportunity for localities to reflect on themselves, and are also lessons can be drawn upon, and based on which, more problems and weak lines are identified in current work. Then, corresponding steps were taken to mitigate these gaps and improve the relevant systems. For example, authorities of Guangdong Province have stepped up water pollution control. In 2017, Shenzhen city built approximately 2,000 km of pipe network for sewage treatment, equaling the total of that built in the last 5 years combined.

Fourth, accountability in environmental protection inspection was further strengthened. It's quite normal right now that people who failed their duties in protecting ecological environment are held liable, which is also a part of our work. The signal sent to official is almost clear that so long as one fails his environmental protection duties, punishment will surely be instituted, in a view to urge officials to fulfill their due responsibilities. In particular, the mechanism to hand over the inspected cases to localities and urge them for further investigation and rectification is working quite well, which has actually put pressures on the localities and acted as a deterrence discouraging from inaction or malpractice.

Fifth, traditional development mentality was transformed. Driven by environmental protection inspection, localities took the initiative to adjust the development mode. While fixing the found problems, they built up the momentum and went one step further to adjust industrial structure and energy mix, which helped to transform the industries and greatly advanced the supply-side structural reform. For example, new real estate projects are no longer allowed in the core ecological conservation area of Hainan Province. Zhejiang Province is now riding on the tide to upgrade its traditional industries. With tangible results achieved in so many localities, it's reasonable to say that inspection is never a one-off deal. Thank you.

Beijing Evening: I have a question might off today's topic. Could you please update on the latest research findings of National Joint Research Center on Air Pollution Prevention and Control on the causes for heavy air pollution? What work has been done and what's the role of it?

Liu Youbin: In this September, the MEP officially launched the project on cause analysis and control of heavy air pollution, funded by the National Science Fund for Distinguished Young Scholars under the strong support of our Premier Li Keqiang. Experts involved in this project had on-site research in the key areas, joining the localities in analyzing sources of fine particulates, producing emission inventory for major sources of air pollution, revising the contingency plan for addressing heavy air pollution weather incidents, and working out plans for comprehensive control of air pollution, in an effort to support local decision-making on air pollution prevention and control. So far, all of the 28 sub-studies under the 4 major thematic studies of the project have been launched, and a large-scale, all-dimensional and integrated network for observing weather incidents of heavy pollution is set to be established. With the research unfolding on all fronts, we are able to control air pollution in a more science-based and detail-oriented manner, thus paving way for a sustained improvement of air quality in Beijing–Tianjin–Hebei Region in the autumn and winter seasons.

By research, experts got a deeper understanding of the sources of and causes for heave air pollution in autumn and winter seasons. They drew the conclusion that coal burning, automobile, and industrial production are the three major causes for heavy $PM_{2.5}$ pollution in the Beijing–Tianjin–Hebei region and its periphery in the autumn and winter seasons. Of the three, coal burning is the prime cause, contributing to 50% of $PM_{2.5}$ concentration in the heavy polluting seasons. We analyzed the recent weather incidents of heavy pollution since this autumn, and found that the share of sulfate in the $PM_{2.5}$ concentration in Beijing was brought down to around 10% and even lower, indicating ultralow retrofitting in thermal power industries and regulation over companies in unauthorized locations, lacked the proper certificates and failed to meet emissions standards are making major headways. Yet at the same time, the share of nitrate increased to above 30%, suggesting that NOx emitted from automobiles are increasingly contributing to this growing share, and thus have to be further regulated.

Research findings also indicate that surge of $PM_{2.5}$ is mainly attributed to local emission, regional transport, and increase of secondary PM. The impacts of the three factors vary based on locations and time periods, and usually interact with each other. In cities like Shijiazhuang, Tangshan, and Handan with intensified pollutant emission, usually the first factor would take hold when meteorological conditions worsened. The air mass with high $PM_{2.5}$ concentration then moves to the downwind cities driven by prevailing winds, eventually getting more violent when interacting with the local pollutants and creating more intense pollution. Gaseous pollutants such as sulfur dioxide, nitrogen oxides, and volatile organic compounds evolved under suitable conditions into secondary particles such as sulfates, nitrates, and secondary organic aerosols, and further exacerbated $PM_{2.5}$ pollution due to hygroscopic growth. This is how the third factor brought into force. It's concluded by the experts that surge of $PM_{2.5}$ concentration during autumn and winter seasons are the results of the above three.

Despite the notable effects and progress in controlling and preventing air pollution, it will remain our top priority to address and prevent air pollution during the autumn and winter seasons, which is deemed one of the hardest nuts to crack. Yet with the continued study of experts, as well as the concerted efforts of relevant parties, this challenging problem is on track to be resolved.

Monitoring results show that the share of days with excellent and good air quality for 13 cities in the Beijing–Tianjin–Hebei region rose to 68.5%, an increase of 31.6 percentage points. The $PM_{2.5}$ concentration dropped to 60 μg /m^3, down by 41.2% from last year. But most notably for it is that we have successfully curbed a heavy pollution weather incident days before. It began on Nov. 4, when we activated the orange alert for heavy pollution weather. Thanks to the immediate, collaborative and targeted measures taken by concerned parties in reducing emission, the impacts of heavy pollution were drastically brought down with 20% reduction of major pollutants emission, lower than the expected peak value, and with relatively short duration. This is a typical case exemplifying how to deal with such weather incident. We also saw media report saying that despite the orange alert, what people saw was still blue and clear sky instead of expected heavy smog. Credit must be given to our targeted measures and scientific understanding of the causes, formation, and rules of the heavy pollution weather incidents.

Such a positive change also shows that measures laid out in the *Action Plan on Air Pollution Prevention and Control* are effective, and we are on the right track and seeing remarkable results in tackling air pollution in autumn and winter seasons. All these fruits are impossible without the work and contribution of the experts involved in the project, which boosted our confidence in winning the battle to defend blue sky.

China Business News: For the Stage I inspection, the inspection teams received and handed over hundreds of thousands of cases reported by the public, addressing a lot number of prominent environmental problems. What are the problems reported by public? What are the features of these problems? It seems some issues remained unresolved for a long time, yet immediately tackled as soon as the inspection team came. Why is the case?

Liu Changgen: It's a major step as well as an important tool for central environmental protection inspection to receive and deal with public complaints. We analyzed the 104,000 cases reported by the public, the conclusions of which are as follows:

First, in terms of spatial distribution, the number of cases reported in eastern region is higher than that of central, so is for the central to the western. It's in average term, given the number also varies in different provinces. Notable for it is that the number in the northeastern China is generally higher than in other regions, above both the national average and the average in east China.

Second, in terms of factors of pollution, we saw most are related to air, taking up around 41%, while that for water is 17%, noise 15%, solid waste 11%, ecological damage 7.5%, and others 8.5%. This is about major factors.

Apart from it, we are more concerned about its relevance with people's interests, and categorize these complaints into the following groups.

The first group is about the seemingly insignificant problems yet concerning people's everyday life. Such cases usually involve kitchen fumes, noise, dust, sewage, garbage and small workshops, which usually cause small-scale pollution, and took up 65%, the largest share, of the total reported ones. There were also a considerable amount of cases involving the rural–urban fringe with poor management, a result of rapid urbanization, where environment was polluted and industrial and residential zones mixed together. These issues on the first look are insignificant, yet in fact affect people's daily life, thus must be seriously dealt with. These are the ones that "remained unresolved for a long time, yet immediately tackled as soon as the inspection team came" as you mentioned. These are environmental problems, as well the urban development ones, which largely depend on the thinkings guiding urban management and the level of it.

The second group is about interregional and cross-watershed problems that concern public interest. Cases like water pollution, ecological damage, and pollution in the surroundings of factories and mines took up around 27% of the total. With one round of inspection after another, the public are more motivated in participating in environmental protection. We are seeing a notable increase in the number of such cases reported by public. Such cases usually involve significant environmental problems that can hardly be addressed in a short term, calling for science-based planning to accelerate the process and a sound interaction between public and the government. There will be spot-checks by the inspection team on the reported cases, and important issues will be included in the final inspection report and handed over to localities for further investigation and punishment.

Some reported cases are about Nimby Effects (not in my backyard), roughly 5% of the total. Though with a small share, these cases usually involve conflicts of interests and over advertising, usually grabbing more public attention. The inspection team will carefully study and take a prudent approach to such issues.

There's a slight 3% on conflicting interests or reports by competitors of the same industries.

It turned out the approach that inspection and rectification proceed simultaneously, which took responding to public complaints as one of the priorities, is quite effective. Honestly speaking, the initial plan on central environmental protection inspection didn't include this approach. It's in fact an approach later developed by our inspectors in accommodating the needs of local people while carrying out pilot inspection in Hebei Province. It now stands out as a distinct feature as well as a major highlight of central environmental protection inspection.

Firstly, this approach is creating a sound inspection environment. It's quite popular today for people to follow, participate in and show support to environmental protection inspection. With the number of reported cases increasing round after round, it's then obvious that our people trust and look forward to central environmental protection inspection.

Secondly, it raises the environmental protection awareness at the community level. The key to putting environmental protection down on ground lies in the grassroots environmental authorities. By proceeding with inspection and rectification simultaneously, issues can be resolved directly and grassroots-level officials also learn more from it.

Thirdly, it helps improve the working style of grassroots-level officials. The process of fixing problems while inspection is taken by localities as an opportunity to improve their working styles. They led the efforts, and strengthen environmental education and accountability, with good results yielded. This echoes what we just mentioned, all those concerning public interests is no small matter, hence central environmental protection inspection is for sure no easy task. Thank you.

The Paper: The MEP has held more admonitory talks with local officials in an even tougher way. What are the major issues talked about and who are talked with? What's the plan on admonitory talks next year?

Liu Changgen: Holding admonitory talks with liable officials is an important tool for MEP to oversee local administrative performance. It yielded very good results, and is highly valued by the Party Members' Group of the MEP. Since this year, the MEP has held admonitory talks with localities where air or water quality continued to degrade, salient problems were identified in the targeted inspection, prominent environmental problems were not effectively controlled, and acute problems found in the follow-up inspection in Beijing–Tianjin–Hebei region on comprehensive air pollution control. Those who are talked with include those from 30 cities, counties and districts, provincial departments, and centrally administered state-owned enterprises, with 6 of them through internal talks, and the rest open talks. This has actually put pressure on localities, propelling them to rectify problems and is also an alert to all.

According to our work plan, next year we will further enhance admonitory talks with liable officials, with focus turned to the following areas.

First, the central government put forward winning the war against pollution, and the recent Central Economic Work Conference also outlined a plan on it. We will act upon this call, having admonitory talks with localities that are acting ineffectively and slowly and found with multiple problems.

Second, such talks will also be conducted with those areas that fail to deliver on the targets set out in the *Action Plans* on air, water, and soil pollution prevention and control and found with notable decline in environmental quality.

Third, those localities that are ineffective in addressing the problems found in the central environmental protection inspection, and have problems surfacing again and leading to negative impacts shall be called for admonitory talks. Apart from it, targeted inspection will be instituted to review the key issues that might cause serious damage, and results will be handed over to localities for further investigation and punishment.

Fourth, localities failing to curb prominent problems that are ordered for rectification by central leaders, disclosed by the media, and of local people's concern are also in the list.

Meanwhile, for some key sectors and regions, a five-step approach will be implemented, which are inspection, handing off cases to local authorities for further investigation and punishment, follow-up inspection to review the found problems and track the progress, admonitory talks with liable officials for environmental problems, and targeted inspection. We will act in accordance with relevant requirements once environmental problems are found.

Admonitory talk, after all, is a means not an end, thus not the more the better. We aim to using it to educate and warn people, and create a deterrence effect. For the localities involved, officials responsible shall be urged to work harder to fix the problems, while for other areas, officials will be warned from making the same mistakes, and act proactively to find out their own problems and try to address them in advance. By which, we hope to amplify the effects of admonitory talks, trying to influence and see positive improvement in a wider range of areas and sectors. That's why we are trying to make admonitory talks more targeted, play an exemplary role, and disclosed to public. Thank you.

The Beijing News: My question is about *Action Plan on Air*. As this year marks the concluding year of implementation of this *Action Plan*, could you share with us the overall progress made so far? Are there any localities yet to deliver on the set targets? What are remaining problems? And what's your plan next?

Liu Youbin: The *Action Plan on Air Pollution Prevention and Control*, which was released by the State Council in 2013, is a milestone document in tackling air pollution. It not only reviewed the progress and lessons learnt in curbing air pollution in the past, but also put forward measures to curb it in the new era. It's only three days away from the conclusion of its implementation, and many are closely following whether we will hand out a perfect answer sheet.

Here are several major targets in terms of improving air quality set forth in this *Action Plan*. PM_{10} concentration in cities at and above prefecture level will drop by 10%, and $PM_{2.5}$ concentration in the Beijing–Tianjin–Hebei region, Yangtze River Economic Belt, and Pearl River Delta respectively by 25%, 20%, and 15%. There's also a target catching a lot of attention, namely the $PM_{2.5}$ concentration in Beijing to be brought down to around 60 μg /m^3.

Since the issuance of this *Action Plan*, remarkable progress was made in the implementation, thanks to the strong leadership of the State Council and collaborative efforts of all parties concerned. I came across several articles recently. They are quite impressive, and some of which were written by you. Some wrote that both the production and the sale of Shawo Radish, a kind of local specialty grown in Tianjin city, saw an increase as a reduced $PM_{2.5}$ allowed more sunlight. Some wrote that air purifiers were overstocked this year. Some others said that with people seeing blue sky more often, it's no longer a surprise to see people posting blue sky in their friends' circle of WeChat.

That's what we saw from the monitoring data. In 2016, the average concentration of PM_{10} in 338 cities at and above prefecture level was down by 15.5% from 2013, and that of $PM_{2.5}$ in the Beijing–Tianjin–Hebei region, Yangtze River Economic Belt and the Pearl River Delta down by more than 30%.

From January to November of 2017, PM_{10} concentration nationwide dropped by 21.5% from 2013, and $PM_{2.5}$ concentration in the Beijing–Tianjin–Hebei region, Yangtze River Economic Belt and the Pearl River Delta were cut by 38.2%, 31.7%, 25.6% respectively, with that of Beijing standing at around 60 μg /m^3.

The *Action Plan* will be concluded in just three days. Same to all of you here, I'm looking forward to seeing the fruits. Analysis of some experts indicated that the goals set forth in this *Action Plan* on improving air quality are able to be attained.

Yet, progress is no excuse for complacency, as we are keenly aware that it's still an arduous task to prevent and control air pollution, and there remains a gap between what we've achieved and that is expected by our people. Here are the three major challenges in air pollution prevention and control. First, it's a prominent problem in some regions with air quality failing to attain the standard in some time. Second, progress varied in different regions with some provinces far lagging behind. Third, the energy mix dominated by coal, the industrial structure by heavy chemical industry, and the transport structure by road transport were all yet to be changed, and generated a large amount of pollutants.

Air pollution prevention and control is now standing on a new starting point. Looking ahead, we must not slacken our efforts or lower our guard. In accordance with the arrangement and requirements of the 19th National Congress of the CPC, we will remain truc to our original aspiration and keep our mission firmly in mind, continuing with air pollution prevention and control to defend our blue sky. Firstly, we will continue to implement the *Action Plan in Combating Air Pollution in Beijing–Tianjin–Hebei Region and its Periphery during the Autumn and Winter Seasons for 2017–2018*, trying to kick off a good start in the fight for blue sky. Secondly, we will ask the China Academy of Engineering to evaluate the implementation of the *Action Plan on Air Pollution Prevention and Control*, through which lessons will be learnt, shortfalls identified and measures put forward for further improvement. Thirdly, we will study and draw up guidelines for the phase II action plan on air pollution prevention and control to make clear the timetable and roadmap of our fight for blue sky. Thank you.

Xinhua News Agency: The total 6 regional inspection bureaus of MEP have been upgraded from public institutions to administrative organs, then how do we align their responsibilities with the central environmental inspection system? And what are the new requirements for their work?

Liu Changgen: Recently the 6 regional inspection bureaus of MEP are operating under the new name of "bureau" instead of "center", indicating a shift from a public institution to an administrative organ. Previously the staff were allowed to enjoy the same treatment with those from an administrative organ, but still had a very different status. The shift shows the high importance attached by the CPC Central Committee and the State Council to ecological civilization and environmental protection, particularly to the environmental inspection.

Over the past 2 years, staff with the 6 bureaus have well performed their inspection duties, took solid steps and exhibited a firm resolve, rising as the mainstay of central environmental inspection. They were ready to respond to calls at all times, and fight hard to complete their missions. This reform has realized two shifts, including one from "center", a public institution, to a "bureau", an administrative organ, and the other from inspection of violation cases only to examining both violation cases and people liable for such cases. All these helped the inspectors better fulfill the duties of central environmental protection inspection. Its significance is mainly reflected in the following two aspects.

First, more focus is given to administrative supervision. Previously, the focus of the 6 regional inspection bureaus is mainly on supervising and examining how enterprises are dealing with the environment, and local governments' performance in enforcing national environmental protection policies, regulations and standards. Generally speaking, they were mainly inspecting the violation cases. The reform, however, explicitly requires that they not only undertake central environmental protection inspection, but also examine officials liable for such cases to ensure both cases and people liable are brought to light. It's an integral part of central environmental protection inspection.

Second, it facilitates the establishment of a more professional inspection team. Environmental protection inspection is expected to be more challenging in the future, articulating the needs for the inspectors to learn more and practice more to better inspect both violations cases and liable officials. With the reform further clarifying the roles and functions, the inspectors will commit themselves to further learning and practicing environmental protection inspection, and work to build a more professional team through continuous practice, so as to better take on the inspection tasks. Thank you.

Reuters: As we are entering the winter heating season, there was media coverage from time to time on natural gas shortage or irregular gas supply in some villages and communities across the Beijing–Tianjin–Hebei Region. What's the response of MEP?

Liu Youbin: People are very concerned about the coal-to-gas project and clean heating in north China. In advancing coal-to-gas project and clean heating, the MEP has always been putting people first, working hard to ensure a warm winter for all, and will seriously deal with gas shortage that delays winter heating.

On December 4, the MEP circulated an extra urgent document to the 2+26 cities across the Beijing–Tianjin–Hebei Region and its periphery, namely the *Circular on Meeting People's Heating Needs While Managing Small-scale "Scattered" Coal Burning*. It made clear that the top priority was to guaranteeing people a warm winter, and also provided that during the heating season, localities yet to finish the coal-to-gas and coal-to-electricity conversion would be allowed to follow the traditional coal heating or take other alternative heating methods.

In December 15 to 20, the MEP organized over 800 survey and inspection teams comprised of more than 2,000 people seconded from relevant departments and bureaus within the MEP or its affiliated institutions for a tour around the Beijing–Tianjin–Hebei Region to examine the actual heating conditions there and possible difficulties of local people may face in getting heat. Minister Li Ganjie, Vice Minister Zhai Qing and Zhao Yingmin all led teams to the local communities and to each household for first-hand knowledge of their actual heating conditions. In 6 days, over 5 million households of 25,220 villages in 2,590 counties (towns, streets) of 385 prefectures (cities, districts) were examined.

On-site inspection found out that some villages and communities finished with coal-to-gas and coal-to-electricity projects did suffer from gas shortage in a time, but mostly were eased later by increasing gas supply and referring to other temporary measures like coal burning and electric heaters before we had arrived. Where heating remains a problem in our inspection, our inspection team would be there until problems were solved. By the night of December 20, heating needs for all concerned local residents were well met with joint efforts of relevant departments and local governments.

In the inspection tour, we heard many local residents saying that thanks to the coal-to-gas and coal-to-electricity projects, they don't have to get up at midnight stoking the furnace to keep warm. Their house got warmer and cleaner, and their quality of life significantly improved. The local governments also took the project as one concerning people's well-being, and one to the exact need of our people. Some experts and scholars also shared their views. Mr. Yao Qiang, head of Graduate Education, and former head of Institute of Thermal Engineering with Tsinghua University, Mr. Hao Jiming, Academician of the Chinese Academy of Engineering, and Mr. Lin Boqiang, Professor with the China Institute for Studies in Energy Policy of Xiamen University, all pointed out that an important piece of experience we learnt from developed countries in addressing heavy air pollution is to use clean energy, which was also in line with our national conditions when dealing with air pollution. We were on the right track and the results were notable. It's a historical leapfrog to free people from burning coals and firewood. Some energy experts also analyzed the reasons for gas shortage in some localities this year, concluding that coal-to-gas is by no means the prime cause for it.

The Party and the central government gave high priority to replacing coals with clean energy in North China. On December 21, 2016, President Xi Jinping noted in the 14th meeting of Central Leading Group for Financial and Economic Affairs that advancing clean heating is critical to ensuring a warm winter for the people, as well as to reducing haze and smog weathers, and it is an important step toward optimizing energy production, advancing consumption reform, and transforming lifestyle in rural areas. The government report early this year specified that we shall implement coal-to-gas and coal-to-electricity projects in the 2+26 cities, which is now advancing smoothly and realizing both social and environmental benefits. Statistics show that $PM_{2.5}$ concentration in Beijing–Tianjin–Hebei region stood at 60 $\mu g/m^3$ in November, down by 41.2% from last year, and the share of days with excellent and good air quality was up by 31.6%.

Clean heating and a warm winter is by no means an either-or choice. They go hand in hand instead of against each other. Coal-to-gas and coal-to-electricity conversion are intended to improve air quality by using alternative energy other than coal, and as well an important move toward improving people's living standard in urban and rural areas. In this connection, this is also a project concerning people's livelihood. Recently, 10 ministries and commissions including the NDRC, National Energy Administration, the MOF, and the MEP jointly issued the *Winter Clean Heating Plan in North China (2017–2021)*. It stipulated that by 2019, clean heating rate in North China must reach 50%, and of which, that in the 2+26 cities at least 90%, prefectures and urban–rural fringe above 70%, and rural areas more than 40%, setting the course for future clean heating efforts in North China.

We believe that with concerted efforts of the entire society, we will well implement the coal-to-gas and coal-to-electricity projects and deliver multiple benefits, contributing to the improvement of both air quality and people's livelihood.

Financial Times: My first question is that how many enterprises were shutdown and banned from production for failure of compliance, and how many had finished corrective steps as required? Second, following the pilot inspection in Hebei Province at the end of 2015, we saw an even tougher inspection instituted this year. Why is that? And what are the improvements in the recent several rounds of inspection compared with the pilots in 2015? Last, will a Stage II and III inspection be launched in the future, and will still in the same frequency and intensity as in Stage I?

Liu Changgen: Thank you. We ourselves also made some analysis on issues of enterprise shutdown and suspension of production you just mentioned. There were indeed a few enterprises ordered to do so, which was partly a result of the one-size-for-all approach of local governments. And in this August and September particular, we heard a lot more criticism from public on it. I think the following two reasons can explain the widespread complaints about one-size-for-all approach.

First, as we are advancing environmental protection inspection, its deterrence and warning effect is amplifying. Right before or after the inspection team arrived, a few local authorities, particularly the grassroots-level party committees and governments, for fear of being called to account, directly ordered the enterprises to suspend production and shutdown irrespective of the actual conditions. It interrupted people's normal life, and ruined the reputation of central environmental protection inspection. However, fortunately, we only saw this in a few individual cases of certain localities and it's only a temporary phenomenon. Yet at the same time, it's no denying that there is typical bureaucratic inertia leading to the one-size-for-all problem, which is not allowed and will be resolutely cracked down on. We also require localities to examine themselves by drawing upon others' lessons, and seriously deal with similar problems found within their own jurisdiction.

Second, as you just mentioned, with the central environmental protection inspection unfolding in full swing, its deterrence effect is further strengthened, and further complicated the intertwined interests. Some resorted to the "we media" like MicroBlog and WeChat to exaggerate the effects and mislead the public. Our leaders were quite concerned about it, particularly instructing us to investigate those cases. We examined them one by one, and also called on localities to find out the truth, clarify the results to the public, and hold those to account for misleading public. For example, in the third round of inspection, some reported that price for paper products was surging as a result of shutting down paper-making enterprises for environmental reasons. However, the investigation results showed that only 10 out of the 3600 paper-making enterprises were closed for violation of environmental laws, far from impacting the overall price.

We did shutdown a few enterprises, and based on our analysis, these were also the ones strongly opposed by the public and failing to fix the problems. There were good reasons for us to do so and we do this in strict accordance with the law. However, there were also a few forced to be closed without giving a reason, which brought great inconvenience to people's life. Corrective actions were requested and taken when we were inspecting them.

You also mentioned a tougher inspection. I think it's mostly associated with different perceptions in different stages, as it takes time for the public to truly understand what the central environmental protection inspection is all about. At the very beginning in Hebei Province, as our pilot inspection only touched upon a very few sectors, plus that the public were still doubting how much impact it might bring about, they didn't realize the potential urgency or pressure from it, so there's the perception that inspection is not tough. In the last several rounds of inspection, accountability system went into greater depth, passing on the pressure to the regulated entity, and they felt the pressure and understood the inspection better. On our part, we have from the start taken consistent measures to it through the first to fourth round of inspection, always been problem-oriented and accountability-focused.

We have always been improving our work mode with more inspections carried out. We are stepping up publicity and reports on inspection, and increasing relevant reporting on TVs, party newspaper and government websites at provincial level. We learnt from those experience that environmental protection inspection requires the enabling atmosphere and public participation, and that's why we laid particular emphasis on information disclosure and publicity both in institutional design and at the operational level. When inspecting localities, the team usually requires the localities to make public relevant information in the format of a list or a chart, and have the progress in rectifying problems reported through local TVs. This has actually created an enabling atmosphere for inspection, urging localities to fix and root out the problems with double efforts. It also encouraged public participation. In light of the above, I think it's already clear that what's behind the wide perception that environmental protection inspection is getting tougher and tougher is the enabling atmosphere we created.

After the inspection, localities are also required to disclose more relevant information and report the progress in addressing the found problems, for example, weekly reports through the media platforms of TV, party newspaper and government website. This may include some inspection tours conducted by party and government officials to urge relevant efforts, progress made in overhauling problems, and to what extent the sense of fulfillment delivered to the people, as well as cases dealt with. All these are aimed at keeping up the momentum of a strengthened inspection and urging corrective efforts. If one locality makes public relevant information through TV, party newspaper and government website, there will bring peer pressure on others. If one reports the information, while another not, the latter will certainly try to do better for not lagging behind. It turned out to be a very effective measure. Many local officials, particularly those from the environmental protection institutions, had feedback that information disclosure and publicity actually gave to light the paramount significance of environmental protection, pushed it to a new high, and also drove local authorities to better fix the problems.

In addition, all concerned information is made to public in a timely fashion, including when inspection is launched and the team arrives, what are the feedbacks to the localities and any progress in their fixing the problems, as well as progress in addressing noncompliance in inspection and relevant accountability. In particular, people can access all relevant information on the main content of inspection reports, compliance plans and progress with it through mainstream media of the central government and the major media outlets of localities. We put all these in the sun to leave it to the public to make the judgement, which represents a huge pressure to the local government and put an end to the one-way "top-down" supervision model.

As for next steps, we are discussing specific steps with concerning departments. The report of 19th National Congress of the CPC had a comprehensive review of and outlined plans on advancing ecological civilization and environmental protection. It put forward specific tasks and requirements in terms of promoting green development, solving prominent environmental problems, intensifying the protection of ecosystems, and reforming the environmental regulation system. The recent Central Economic Work Conference also made comprehensive plans on winning the war against pollution. In the future, we will act in accordance with all these plans and requirements, and the focus are set as follows:

First, we will have a systematic review and analysis of the Stage I central environmental protection inspection, review the experience, identify the problems, and improve the relevant mechanisms accordingly, to be well prepared for the upcoming inspections.

Second, we will work to advance legislations and institutions on environmental protection inspection, and further refine relevant mechanisms, to make environmental protection inspection more law-based and standardized.

Third, we are studying to launch another round of inspection to review the found problems in the Stage I inspection and track its progress. These problems will be closely followed and accountability will be further strengthened to ensure due efforts are in place. Meanwhile, we will also initiate a targeted inspection on prominent environmental problems affecting people's daily life on an irregular basis, with extra attention paid to air pollution in key regions, black and putrid waters in key cities, as well as that are highly relevant with the priorities set in the fight against pollution. This will greatly facilitate the national efforts to win the war against pollution.

Fourth, we will actively guide the localities in setting up a provincial environmental protection inspection system, to realize a 2-tiered inspection regime with one at the national level to ensure provincial compliance, and the other at the provincial level to inspect cases and hold accountable officials that are at the municipal and county level. This will further promote joint actions at various levels.

Above is the preliminary plan. Once it's agreed upon by the relevant central departments, we will submit it to the CPC Central Committee and the State Council for review and approval. Thank you.

Beijing Youth Daily: With 2-year efforts, the central environmental protection inspection realized a full coverage of the 31 provinces, autonomous regions, and municipalities. What impressed you most in the whole process through its formulation to the implementation and to addressing noncompliance and strengthening accountability? Could you share with us your feelings?

Liu Changgen: As you mentioned, we do have mixed feelings about it. In the 2-year inspection, we were encouraged a lot while at the same time bore heavy pressure. Minister Li Ganjie also elaborated on the inspection when taking questions from the press in the 19th National Congress of the CPC. He highlighted the following 6 points that contribute to the success of central environmental protection inspection. First, we are committed to making development people-centered, and seriously took the problems of public concern and acted on solving them. Second, we strengthened the consciousness of the need to maintain political integrity, think in big-picture terms, follow the leadership core and keep in alignment, taking a clear political stand. Third, the inspection on party and government officials was strengthened to ensure that they are held equally responsible and leading officials must take responsibility for environmental protection within their jurisdiction. Fourth, we followed a problem-oriented approach. Fifth, information was disclosed timely for the public to oversee. We made public every reporting received. Sixth, those officials responsible for environmental damage were held accountable. These are exactly to the point, and I also felt the same as one deeply involved in this process. Here I'd like to share with you some of my observations.

First, the CPC Central Committee with president Xi Jinping at its core exercised full and rigorous governance over the party, and improved party conduct and enforced party discipline, fostering a healthy political atmosphere of integrity within the party that underpinned central environmental protection inspection. I've been engaged in environmental protection inspection for 7 to 8 years, and am keenly aware that it's crucial to ensure a full and rigorous governance over the party, improve party conduct and enforce party discipline, as well as to foster a healthy political atmosphere of integrity within the party. Without which, orders wouldn't be implemented in full, and central environmental protection inspection would go nowhere. Central environmental inspection is designed to confront the tough with toughness, and will not only examine the work performance of local party committee and government in protecting the environment, but also their political will in acting upon the CPC Central Committee's decisions and plans. It's with this healthy political atmosphere of integrity, can central environmental inspection be advanced smoothly and grow from strength to strength.

Second, science-based decision-making, strong leadership and strategic resolve of the CPC Central Committee and the State Council lie at the core that drives forward environmental protection inspection. I've been engaged in and coordinating environmental protection inspection, thus understanding deeply the great political advantage of socialism brought about by the CPC Central Committee in advancing environmental protection. The central environmental inspection plan has very clear goals and very specific steps, making it highly implementable. It also called on local party committee and government to take responsibility of environmental protection within their jurisdiction, which is very essential for future environmental protection actions. Certainly, side by side there are also pressure and backlash impeding inspection. Some localities at the beginning greatly doubted the effectiveness of a problem-oriented approach, and some were reluctant to hold to account liable officials. In particular, some enterprises and individuals negatively affected by our inspection misled the public by intentionally exaggerating the negative effects of environmental protection on economy and commodity price, which also dwindled the effects of environmental protection. However, the CPC Central Committee and State Council displayed strong leadership and strategic resolve, encouraging and supporting us as always. President Xi Jinping made important instructions on each milestone, which boosted our confidence and was also ensuring the inspection was stepping up and on right track.

Third, the public trust us, actively engage in the process, and provide wholehearted support. It encouraged us to keep on, and also made the inspection more influential and penetrating. In the pilot inspection in Hebei Province, we received a lot number of calls and letters from people, showing that the public were expecting us to address environmental issues in the inspection. Honestly speaking, we were very pressured and sensed a heavy burden on our shoulder. That's why we worked hard and made every effort to address public complaints. At the very beginning when we haven't yet published the feedback report, many colleagues were questioning why central environmental protection only focused on those trivial issues. In our view, these issues brought forward by our people should and must be properly addressed, so that the people would have a strong sense of fulfillment. With the wholehearted support of the people, we are more confident in meeting those challenges. In the last four rounds of inspection, we strengthened information disclosure, working hard to address noncompliance during the inspection, which further motivated public participation. With active public participation, recognition, and trust, we are able to hear the voice of our people and make our efforts more targeted. This further enriched the central environmental protection inspection, and provided a solid foundation for it. For your information, each round of inspection had received many letters from the public before they were leaving. Some expressed their gratitude, which came in various forms, even poems and songs, to President Xi Jinping and the CPC Central Committee and the State Council for taking concrete steps and acting promptly to address immediate public concerns. We were deeply touched.

In addition, as an inspector, I fully understood that as central environmental protection inspection is a very high-level one, involves a spectrum of areas, and usually touches upon very sensitive issues, it's crucial to conduct inspection in a standard and serious manner and ensure its authoritativeness. These also constitute the basic requirements of environmental protection inspection, and if failed to meet them, inspection can hardly be successful, and that's why we have been working hard to meet them. For example, drawing upon previous experience, we came up with a series of basic procedures and regulations guiding inspection teams to the designated localities and carry out inspection, as well as other disciplines and rules for them to follow. We also proposed 7 key steps for conducting inspection, which are preparation, arriving at the designated locality, drafting inspection report, providing feedback, handing over cases to local authorities or other departments, urging corrective steps and filing for record, and developed specific guidance for each step. We set in place work mechanisms to guide the inspectors' work after they arrived the designated locality, including collaborative action and shared duties, exchange and communication, handing over cases to relevant authorities, overseeing corrective steps, holding regular meetings, and guiding public opinions. We also proposed basic inspection tools, including mobilizing and training the inspection teams, requesting relevant information and documents, holding individual talks, visiting and interviewing local people or concerned parties, receiving and dealing with public complaints, and conducting on-site inspections to localities. There were also more than 50 inspection systems, models, and paradigms available to facilitate interviews with local people or concerned parties, collecting evidence, and addressing noncompliance during inspection. All these contribute to a standard and systematic inspection.

Above are some of my observations. There are also other important matters that require further attention in inspection such as how to exchange ideas with localities, win their understanding and support, and ensure violation case brought to light and people liable held to account. Due to time limit, I will not go into details. Thank you again.

Liu Youbin: Today's regular press conference marks the 12th of its kind as well as the last one of this year. Over the past year, friends from the media greatly supported our work, not only by your timely coverage, but also the constructive ideas and suggestions proposed to us. As such, our work has been constantly improved.

2017 marked a remarkable year for China's environmental protection. Friends from the media actively reported progress of China's environmental protection, sharing the country's wonderful stories of environmental protection and editing which into excellent news. You also worked diligently to report almost each of the MEP's thematic publicity event including the "central environmental protection inspection," "protection of drinking water source areas of the Yangtze River Economic Belt," "combating weather incidents of heavy pollution," "follow-up inspection to review the found problems," "celebration of World Environment Day," "war against air pollution," and "battle to defend blue sky," to name just a few. You are the witness and recorder of China's environmental protection, as well as the participant and contributor of it. We would like to take this opportunity to express our heartfelt thanks and regard to your productive work over the year.

With another 3 days, we will embrace the new year of 2018. Wish you all a happy new year.

That's the end of today's press conference. Thank you.

Part III
Special Press Conferences of MEP

Press Conference on MEP's Stepped-Up Action for Comprehensive Air Pollution Prevention and Control in the Autumn and Winter of the Beijing–Tianjin–Hebei Region and Its Periphery

September 1, 2017

On the morning of September 1, MEP held a special press conference to introduce the stepped-up action for comprehensive air pollution prevention and control in autumn and winter of the Beijing–Tianjin–Hebei region and its periphery. The press conference was attended by Liu Bingjiang, Director General of the Department of Air Environment Management of MEP, Tian Weiyong, Director General of the Environmental Supervision Bureau of MEP, and Liu Changgen, Deputy Director of National Environmental Protection Inspection Office.

Ministry of Environmental Protection of the People's Republic of China,
2017 Press Conference Records of Ministry of Environmental Protection, the People's Republic of China, https://doi.org/10.1007/978-981-13-7330-5_14

Press conference of MEP's stepped-up action for comprehensive air pollution prevention and control in autumn and winter of the Beijing-Tianjin-Hebei region and its periphery (1)

Press conference of MEP's stepped-up action for comprehensive air pollution prevention and control in autumn and winter of the Beijing-Tianjin-Hebei region and its periphery (2)

Liu Youbin (Moderator): Friends from the media, good morning! Welcome to today's press conference.

The year of 2017 is the final year of the first phase of the *Action Plan on Air*. To do everything, we can to control air pollution in autumn and winter and properly deal with heavy pollution weather is our top priority at present. To prevent and control air pollution in autumn and winter, improve the environment, solve environmental problems affecting people and speedup the transformation and upgrading of the economic structure, MEP, together with the National Development and Reform Commission, the Ministry of Industry and Information Technology, the Ministry of Public Security, the Ministry of Finance, the Ministry of Housing and Urban–Rural Development, the Ministry of Transport, the State Administration for Industry and Commerce, the General Administration of Quality Supervision, Inspection and Quarantine, National Energy Administration and the people's governments of six provinces and municipalities in the Beijing–Tianjin–Hebei region and its periphery, decided to step up efforts for the comprehensively controlling autumn and winter air pollution 2017 and 2018.

The stepped-up efforts will be geared toward problem-solving. Successfully controlling the enterprises and clusters that are in unauthorized locations, lack the proper certificates and fail to meet emissions standards (hereinafter referred to as enterprises with "location, certificate and emission" issues) and making the emission from pollution sources high above the ground persistently meet the standards are our purposes, for which our main approaches will be consuming less coal, elevating standards and upgrading emitters, and conducting off-peak production. The key will be properly dealing with heavy pollution weather by strengthening joint prevention and control, implementing laws and regulations strictly, bolstering inspection, accountability, and transparency and mobilizing all people to engage. For that, the "2+26" cities on the pollution transmission channels in the Beijing–Tianjin–Hebei region will be involved. The main target is to meet the assessment indicators specified in the *Action Plan on Air*, which are an over 15% reduction of the average $PM_{2.5}$ concentration and of heavy pollution days over the previous year in the transmission channel cities.

Ten departments and ministries including MEP and 6 provinces and municipalities formulated an overall plan for the stepped-up action to ensure that its targets will be met, which involve 32 key tasks on 11 fronts and 9 safeguard measures. Also, 6 supporting plans were made to step up inspection, follow-up inspection, targeted inspection, quantified accounting, information disclosure, and publicity, a systemic portfolio arrangement for the stepped-up action. A long-term mechanism will be established to allow us to fulfill each task with great determination one by one until the blue sky prevails over pollution. On August 31, Minister Li Ganjin chaired the work meeting to make comprehensive mobilization and arrangements for implementing the stepped-up action.

We are joined today by Liu Bingjiang, Director General of the Department of Air Environment Management of MEP, Tian Weiyong, Director General of the Bureau of Environmental Supervision of MEP, and Liu Changgen, Deputy Director of National Environmental Protection Inspection Office.

The floor is open now.

China National Radio: MEP has been stepping up pollution control in recent years and takes action every winter. What is the overall arrangement for the stepped-up air pollution control efforts this year? What are the new highlights? Is it possible to avoid the frequent occurrence of heavy pollution weather, as was the case last winter, through the stepped-up action?

Liu Bingjiang: First, as you said, in recent years MEP has taken a lot of measures to prevent and control air pollution. The stepped-up action plan for air pollution in the Beijing–Tianjin–Hebei region and its periphery is the first that targets autumn and winter. In previous years we formulated only annual plans.

As you can see, heavy pollution weather occurs frequently in autumn and winter, the drop in the concentration of $PM_{2.5}$ over the past four years is limited, and the number of days with heavy pollution remains basically the same. In light of this, we organized experts, local governments and ministries and commissions to carry out a thorough research. This year we, for the first time, set specific targets for autumn and winter: a 15% drop in both the average overall concentration of $PM_{2.5}$ and the number of heavy pollution days in autumn and winter. The targets set for cities are not the same, being higher in key cities.

Second, this is the first time that targets have been quantified for the "2+26" cities. The targets set previously are mostly general ones, while this time, the target is broken down into quantified city-specific sub-targets and tasks has also been specified at the county, district, township, and village level. This is a prominent feature of this year. The stepped-up action plan for autumn and winter is necessary, justified, and urgent, and most importantly it addresses air pollution. But it is not just about air quality; it is also about supply-side structural reform and industrial transformation and upgrading. What we want is to improve air quality and at the same time facilitate industrial transformation and upgrading.

Liu Bingjiang, Director General of the Department of Air Environment Management of MEP

The problem with the previous top-level design is that it was not well implemented. Therefore, this time, along with the stepped-up action plan, we also formulated six supporting documents as a portfolio of moves. In addition, eight–ten cities will be selected as targets of very rigorous central special environmental inspection. Also, we will have a 5,600-strong reinforced inspection team. In autumn and winter, competent government staff will be deployed in the "2+26" cities, joining the locals in their stepped-up action. So you can see that the overall plan is specific and quantified, including the regulation and accountability part. This time, neither the top-level design nor enforcement will be faulty. We will be problem-oriented and fulfill the quantified targets one by one. This is a process that localities must abide by. However, localities should view the quality target as the ultimate end as the process is one means that does not necessarily deliver the ultimate end. When necessary, local governments should also take their own measures depending on the degree of quality improvement. We equipped every district and county with air quality monitoring stations, which will contribute to more precise monitoring and assessment and by extension more precise emission-reducing measures, regulation and accountability. This is another prominent feature of this year.

You are all concerned about whether this winter will be like the last one. Well, that is closely related to meteorological conditions. Anyway, this year's target of air pollutant emission reduction is twice that of last year, nearly thrice. In addition, compared with last year, the emphasis and pressure on implementation are greater this year. Based on the two reasons, I believe this year we will achieve a big emission reduction. Let me give you an example. Last year, 800,000 households replaced coal with electricity and natural gas, this year the target is three million households; last year, less than 10,000 small coal-fired boilers were shut down or suspended, this year the target is 44,000. All ministries and commissions are very cooperative in both pricing policies and funding. For example, clean heating pilots funded by state revenue have been launched in 12 cities. Regarding coal quality, all ministries and commissions will be very cooperative. Our emission target this year is two to three times that of last year, representing a big increase. What will this year's winter be like? Cold or warm? The *Action Plan on Air* took effect in 2013 and since then went through the toughest 4 years as 2014 had a warm winter, 2015 warmer, and 2016 the warmest of all. Many people hope not to have another warm winter this year. Let me give you some other figures. In the case of Beijing, from 1949 when the new China was founded in 2012, air stagnation that lasted a week occurred only five times. However, it occurred seven times in 2013 alone, and six times in 2014, 2015, and 2016, respectively, and four times in the first half of this year. We do not yet know its occurrence in the next half of this year. But we will organize the Institute of Atmospheric Physics to discuss the matter on September 5 and we may then know whether it is a warm or cold winter ahead. This year should be better than last year, in that I have confidence.

The Papers: It has come to our attention that the stepped-up action includes rules of "quantified accountability" for local governments, can you elaborate on it? Now we face some issues, for example, the remediation of enterprises with "location, certificate and emission" issues is insufficient; the replacement of coal by electricity and natural gas is tarnished with dishonesty; not all small coal-fired boilers have been shut down or suspended; the off-peak production in key industries has not been well implemented. Why does quantified accountability pay more attention to these issues?

Liu Changgen: The document on quantified accountability is a major document on exerting pressure on lower level governments in the "1+6" set ("1" refers to *Action Plan for Comprehensive Air Pollution Prevention and Control in the Autumn and Winter of the Beijing–Tianjin–Hebei Region and Its Periphery in 2017*, and "6" means the six supporting plans). Since the beginning of this year, the Beijing–Tianjin–Hebei region and its periphery has secured good results in air pollution control, but the atmospheric environment situation is still very grim. To effectively exert pressure on localities and consolidate accountability, MEP's Leading Party Members' Group, in accordance with the requirements in relevant central government documents and the guiding principles of the 10th meeting of the air pollution prevention and control coordination group of the Beijing–Tianjin–Hebei region and its periphery, issued regulations on quantitative accountability. The main considerations are as follows:

First, a sword of Damocles will hang over localities to urge them to truly devote themselves to work. During the process of supervision, inspection and law enforcement, we feel that governments and party committees at all levels, generally speaking, have been working very hard and accomplished a great deal. However, in some localities, particularly at the community level, they have not fully fulfilled their responsibility in environmental protection and have been quite passive. Therefore, the focus of quantitative accountability is the leaders and cadres of governments and party committees at the county, district and prefecture level and in relevant departments of the "2+26" cities. The focus of this document is accountability. At present, the situation of air pollution prevention and control in the Beijing–Tianjin–Hebei region and its periphery is grim, especially in autumn and winter. In the course of air pollution control, perfunctoriness, slacking, inaction and abuse of power are not allowed. Anyone that does not act may face consequences and anyone that does not fulfill his due responsibilities may be held accountable. This is what we mean by the sword of Damocles. It will always exert intense pressure and act as a deterrence so that difficulties encountered halfway and at the last mile will all be addressed.

Liu Changgen, Deputy Director of National Environmental Protection Inspection Office

Second, we should pursue a thorough implementation, which is the purpose of quantified accountability that we should work hard to achieve. Success is 10% deployment and 90% implementation. If accountability is not effectively practiced and task assignment is not well implemented, then there will not be much difference between action and inaction, between good and bad performance and between working hard and slacking off. In that case, our plan will be just a plan on paper and remain futile. In contrast, quantified accountability is very specific in all aspects, including what are the matters to be addressed, who will be held accountable and to what a degree will accountability be pursued in various scenarios. All these specifics aim at requiring localities to strive for thorough and productive implementation.

Quantified accountability falls into two categories. One is "task-based". If local officials do not rectify the problems spotted by MEP in inspections and follow-up inspections or fulfill the tasks assigned to them, they shall be held accountable. Also, during this process, local officials should learn from the lessons and take the initiative to address other problems. For example, if we still spot many problems after October, we will hold local officials accountable, specifically speaking, the deputy county head if there are 5 problems, the county head if 10, and secretary of county party committee if 15. The second category is "result-oriented". Localities will be ranked based on how they perform in terms of meeting the air quality improvement target, and those that are ranked among the bottom three and do not meet the target will be held accountable. And it is the prefecture-level leaders and cadres that will be held accountable. In this way, the air pollution control tasks are tied to governments and party committees at municipal and county level. Government agencies exert pressure on those at a lower level, who then do the same thing. This can motivate localities to work thoroughly and meticulously.

You just mentioned four issues: the remediation of enterprises with "location, certificate and emission" issues is insufficient; the replacement of coal by electricity and natural gas is tarnished with dishonesty; not all small coal-fired boilers have been shut down or suspended; the off-peak production in key industries has not been well implemented. They are salient problems that affect the air environment quality in the Beijing–Tianjin–Hebei region and its periphery, and they give us the worst headaches, as revealed by supervision and inspection. They need to be addressed by stronger supervision and inspection and by more government efforts. Meanwhile, efforts made to address the four issues also facilitate other tasks, promoting the overall control of air pollution. Such a structural design highlights the keys. And the tasks are well defined, making it easy to emphatically determine whether you have performed them and leaving no room for wrangling.

Third, we will establish a long-term mechanism. The environmental protection requirements that "party and government officials being held equally responsible and leading officials taking responsibility for environmental protection within their jurisdiction" have been made clear by the central government. The mechanism for such requirements and a new work pattern will take shape as long as we repeatedly do well in supervision, inspection, and accountability and exert pressure on governments at lower levels. If the lower level governments are not mobilized properly, many of our requirements will not be materialized.

Yesterday, our minister has elaborated on the accountability issue, so I would like to skip it. Now, I want to talk about our next steps.

First, tightening up enforcement. We will follow the five-step approach of "inspection, assignment, follow-up inspection, admonitory talks and targeted inspection" and hold accountable with no leniency personnel that do not implement tasks assigned, fulfill their due responsibility or take proper action.

Second, being goal-oriented. We will, via implementing strict, specific and quantified accountability measures, nudge governments and agencies at all levels to take effective action. We will ensure that the stepped-up plan will be translated into reality and thus improve the air quality and supply-side structural reform in the Beijing–Tianjin–Hebei region and its periphery.

Third, taking targeted action. We would not try to take care of everything. We just fix our problems. The content and tasks of quantified accountability have all been specified so that we can better lay out the responsibilities of lower level governments and facilitate their work.

Fourth, working with provinces. Should the need arise, we can carry out special inspections with provincial party committees and governments. Regarding quantified accountability, we have made it clear that the materials on it should be handed over to localities and that local party committees and governments will practice accountability, ask for our permission on the result and publicize it for all to see.

CCTV: We thought that the emphasis of air pollution control was placed on big enterprises, why do the enterprises with "location, certificate and emission" issues receive more attention this time?

Tian Weiyong: Recently, people have been wondering why does MEP, which used to focus on big enterprises, pollution sources high above the ground and pollutant emission targets, now turn to enterprises with "location, certificate and emission" issues. The concept of such enterprises was first proposed in the *2017 Air Pollution Prevention and Control Work Plan for the Beijing–Tianjin–Hebei Region and Its Periphery*, and it is confined to certain special industries and fields. The concept is marked by several characteristics. First, it is confined to the "2+26" cities in the Beijing–Tianjin–Hebei region; second, it mainly targets enterprises that emit waste gas without treating it; third, such enterprises are usually plagued with problems like violations of laws and regulations, backward production processes and techniques, excessive pollution and severe pollution of environment. These characteristics are stated explicitly in the abovementioned document.

To address problems caused by enterprises with "location, certificate and emission" issues is necessary for the "Blue Sky" initiative, the supply-side structural reform and, more importantly, the rule of law.

First, from the perspective of the rule of law, we need to comprehensively rectify problems caused by such enterprises. As you all know, the *Environmental Protection Law* and the *Air Pollution Prevention and Prevention Law* clearly stipulate a phase-out mechanism for processes, equipment, and products that seriously pollute the atmospheric environment. The two laws also contain regulations and requirements on stepping up supervision and law enforcement. Enterprises should abide by national laws and regulations, and law enforcement should be fair. Everyone must abide by the law and safeguard the dignity of the law. Everyone is equal before the law. To protect the environment, we must rigorously fulfill our supervisory responsibilities. We oppose to inaction and negligence in the discharge of responsibilities and regulations and incorrect discharge of duties that tries to address the problems we face in an oversimplified manner. Neither of the two approaches is conducive to the rule of law.

Tian Weiyong, Director General of the Bureau of Environmental Supervision of MEP

Second, from the perspective of supply-side structural reform, addressing the problems of such enterprises is necessary. They are not equipped with advanced industrial processes and technologies or pollution treatment facilities, which contributes to their lower cost. They are beyond the scope of regulation, causing the "bad money drives out good" market phenomenon. By stepping up supervision and inspection, we mean that we would urge localities to seriously crack down on such enterprises violating laws and regulations, adjust regional industrial structure and prompt enterprises to upgrade themselves and pursue higher standards, which will usher in a virtuous cycle. Thus, dealing with such enterprises is an intrinsic requirement of supply-side structural reform.

Third, from the perspective of improving air quality, there are two types of industrial pollution sources, the major source (large and medium-sized industrial enterprises) and enterprises with "location, certificate and emission" issues. Thanks to strict law enforcement over the past several years, big enterprises have become much more appreciative of the need to abide by the law. The percentage of major pollution sources meeting the target at the beginning of 2016 was about 70% and rose to near 97% at the end of that year. As for this year, the percentage hovers at around 97% from January to August and tends to be steady.

The enterprises with "location, certificate and emission" issues do not have any treatment facilities, and severely pollute the environment. They were not regulated and thus caused more and more severe pollution, and have become major factors affecting ambient air quality in some localities. Renovation means resolutely phasing out enterprises that are not up to the standard and urging enterprises whose treatment facilities could be upgraded and improved to do so with their facilities. I believe this major adjustment will improve ambient air quality.

Considering the grim environment situation at present, more is expected of our work in such aspects. This year is the final year of the *Action Plan on Air*. Air pollution is a tough environmental bottleneck on our path toward a moderately prosperous society in all aspects. Environmental protection has a bearing on the interests of the vast majority of people and is the biggest contributor to people's well-being, on which I believe we all agree.

China Daily: We all know that the central environmental inspection has achieved "full coverage". This time the stepped-up action again proposed targeted inspection. What is the difference between the two? What are the specific rules on targeted inspection?

Liu Changgen: The central environmental inspection has entered the fourth round, becoming increasingly effective. It draws more and more attention from the general public and media professionals, including you. So why was a targeted inspection initiative launched on top of that? The *Plan of Supervision and Inspection of Environmental Protection (Trial)* approved by the central government stipulates that on top of the routine biennial inspection of all provinces (autonomous regions and municipalities), there can be irregular, random, and targeted special inspection of areas with salient problems. The targeted inspection also has to be approved by the Central Party Committee and State Council and is carried out in the name of the central environmental inspection group. The targeted inspection is geared toward problem-solving and only aims at learning about how party committees and governments fulfill their responsibilities. It is about personnel concerned fulfilling their due responsibilities in tasks assigned to them. This is what the targeted inspection has in common with central environmental inspection. But there are some differences in the following aspects.

First, the scope of inspection. The central environmental inspection is a general inspection that covers the implementation of national environmental protection decision-making, environmental deterioration, and the handling of it, particularly the discharge of environmental responsibilities of local party committees and governments. The targeted inspection, however, is random and straightforward in dealing with specific matters.

Second, the approaches. The central environmental inspectors deal with complaints while targeted inspectors usually do not. Targeted inspectors are mainly tasked with handling specific problems and determining the leaders, managers, supervisors, and executors who are responsible for the problems, leaders in particular. They conduct investigations, interviews, and inquiries to collect evidence, produce a chain of evidence and a file before handing them over to competent accountability authorities.

Third, time arrangement. The central environmental inspectors stay in the inspected province for about a month, preceded by preparations of about 2 months and followed by report drafting, feedback, handover matters, rectification, accountability, and result review. The process is like a TV drama with one episode after another. Basically, this whole process of supervising and inspecting a province takes a year and a half to 2 years, only to be followed by another round of inspection. Targeted inspectors target at specific problems and act fast within a short time span. Its goal is to carry out accountability work in accordance with the law and discipline and based on facts.

Targeted inspection can be an important complement to the central environmental inspection: the former can be initiated any minute in cases of relapse, insufficient rectification and strong public complaint following the departure of the central inspection group. We have actually carried out several targeted inspections and they are very useful in seeing to the rectification dictated by central inspectors.

Let me share with you some general information on the targeted inspection for the "2+26" cities in the Beijing–Tianjin–Hebei region and its periphery. Inspectors may visit the target cities at the end of this year or the beginning of next year and stay for 15–20 days, depending on the specifics, number of problems and workload. A plan has been made and will be submitted to the State Council for its approval before being implemented. That is the overall plan. After their inspection, inspectors will produce a report and an accountability file, which, upon the approval of the central party committee and the State Council, will be handed over to the provincial party committee and government of the target city so that they can take action for accountability. The accountability result shall be approved by MEP and publicized. That is the initial plan.

Reuters: We have noticed some market price fluctuation during central environmental inspections. Some people believe there is a correlation between the two. What do you think about it?

Liu Chenggen: There are two possible scenarios. In the first scenario, localities take advantage of central inspectors' presence to address issues that should have been addressed according to laws and regulations. This is the case in provinces and autonomous regions like Shangdong, Sichuan, Tibet, and Xinjiang who have addressed many problems such as evident violations of laws and regulations, workshops and other pollution sources that should have been shut down or suspended. We encourage this. Of course, we hope that the inspections can help localities develop long-term mechanisms and ramp up their routine work so that they do not have to wait for the arrival of inspectors to act. In the second scenario, party committees and governments of some localities, particularly those at the community level, out of their worry that inspectors might spot problems and hold them accountable, shut down or suspend all enterprises in advance, which leads to short-term and partial undersupply and affects people's life. This is wrong and we strongly object to it.

Prior to the arrival of the fourth batch of inspectors in Chengdu, we learned that some of the restaurants were suspended for no reason and immediately asked inspectors to communicate with the local government over this matter. Chengdu took this seriously and immediately issued a document of clarification, which was widely acclaimed by people.

The first batch of inspectors to Henan province reported that some local governments there muddled through the inspections in a superficial way, which turned out to be the truth. To that, we responded with timely correction measures.

Liu Bingjiang: I would like to add something. Some people like to make everything about environmental protection and command the moral high ground. Price fluctuations of commodities result from supply and demand, among multiple factors. Not everything is caused by central environmental inspections and the stepped-up plan for air pollution control in autumn and winter.

China Youth Daily: Besides $PM_{2.5}$ concentration, people are also concerned about ozone. How is the present ozone pollution in China? What control measures have been taken? Also, the stepped-up action plan emphasizes that the air administration agency for the Beijing–Tianjin–Hebei region and its periphery will be put in place by the end of September? How is it going?

Liu Bingjiang: Let me first answer your second question concerning the Beijing–Tianjin–Hebei region air protection agency, which has received a lot of attention from all sides. According to the central reform arrangements, MEP and the State Commission Office of Public Sectors Reform have formulated the plan for the pilot of a cross-regional environmental protection agency. The plan has been submitted to the central authorities and approved at the 35th conference of the Central Leading Group for Comprehensively Deepening Reform. In the conference, it was made clear that we should sort out the institutions and administration mechanisms to address major problems affecting the atmospheric environment quality and put the plan into action as soon as possible.

The central authorities are deliberating over a reply to the plan, and once that is completed, we expect the plan to be released soon. After the release, we will, in light of the requirements in the reply, waste no time in making proper arrangements. In a word, work on the plan is proceeding orderly.

Regarding ozone, let's first look at the overall situation in 2016. In 71% of the 338 prefecture-level cities in our country, the ozone level hovers around the air quality standard, which is 160 $\mu g/m^3$, a figure basically in line with the international standard. In 2016, 59 cities exceeded the standard, but the percentage of days when ozone level slightly exceeds the standard was only 4.7%, and that of days with ozone level moderately exceeding the standard was lower and that of days severely exceeding the standard even lower. Plus, there was no grave pollution or ozone level higher than the upper limit. The situation in 2016 was overall within control. If you examine the measurements of 74 cities over the past four years, you can see that the overall trend is fluctuation with a slight increase. This is the basic conclusion. Everyone is very concerned about the possibility of photochemical smog in China. The Chinese government takes this seriously. We have closely analyzed all air pollution incidents in developed countries, and experts found out that when photochemical smog occurred in developed countries, ozone concentrations often exceeded 600 $\mu g/m^3$ and even reached 2,000 in some cities. China's ozone concentration is far below that level, thus photochemical smog is impossible now and the possibility of future occurrence is also very small. What is the level of ozone in China compared with that in developed countries? When the concentration of ozone in key areas in China is relatively high, it is basically the same as that in the United States. The average ozone level of the 338 prefecture-level cities in China is about the same as that in the United States about a decade ago. By the end of August this year, for which data is available, the ozone concentration has increased to a certain degree. But we need to factor in the increase of cities. Overall, China's ozone level, despite the fluctuation, is within a reasonable range.

For example, the ozone concentration in Beijing had been rising for four consecutive years until 2016 when it started to decrease, which is what we expect for the next half of this year as well. But all this time, it is still within a reasonable range. Economic recovery and meteorological conditions also have something to do with ozone concentration fluctuation. Via an analysis of China's data available over the past decade, we can see that overall the ozone concentration in the Beijing–Tianjin–Hebei region has been rising somewhat, is stable in the Yangtze River Delta region and basically meets national air quality standard in the Pearl River Delta region. But we take ozone very seriously. After all, China's ozone level overall has been rising, though the ozone concentration in 70% of the cities hovers around the standard. Our focus is on two pollutants that contribute to ozone, namely nitrogen oxide and volatile organic compound. Nitrogen oxide has been a binding indicator for 7 years in a row. The 12th Five-Year Plan required it to drop by 15% and the actual accumulative drop since 2010 is over 20%. Air quality measurement shows that the concentration of nitrogen dioxide has been decreasing slowly but steadily. We also take volatile organic compound seriously. We are on the same page with multiple ministries and commissions and local governments regarding the comprehensive ozone control plan for the 13th Five-Year Plan period. When we solicited your opinions on the plan, local governments were already working on it. We can tell you that volatile organic compound is overall within control. Compared with other pollutants, ozone is relatively easy to prevent and control. If it is 400 $\mu g/m^3$ outdoors, which is very high, it might be only 60 $\mu g/m^3$ indoors. As long as you do not engage in high-intensity exercises outdoors at noon, ozone will do little harm. We do not need to panic.

China Business Network (CBN): MEP has carried out various kinds of activities such as supervisions and inspections to strengthen law enforcement. It has again launched the stepped-up action for air pollution in autumn and winter. So how to establish a long-term mechanism for environmental regulation and law enforcement?

Tian Weiyong: In recent years, MEP has carried out many supervisions and inspections and will carry out more. This is MEP fulfilling its duties, as stipulated by the law. The *Environmental Protection Law* and *Air Pollution Prevention and Control Law* contain many new rules and requirements. For example, as you may know, the enforcement of the *Environmental Protection Law* needs to be published monthly, a task that can be very difficult without information from the supervisions and inspections. It is generally perceived that law enforcement in environmental protection has been increasingly vigorous, particularly since 2015 when the *Environmental Protection Law* took effect. We are earnestly fulfilling all the obligations stipulated in the law with the authority given to us by the law. The government must not do anything unless it is mandated by the law and must do everything mandated by the law. This is what we strive to abide by.

I gather you also worry if these activities will produce only short-term benefits that come and go like a whirlwind. I think such worry is groundless. During all these inspections, be it the central environmental inspection, the year-round 5,600-strong stepped-up inspection team or the stepped-up inspection for autumn and winter this year, we have been exploring some long-term mechanisms. From our stepped-up supervision and inspection of the "2+26" cities over the past several months, we have extracted relevant experience which can be illustrated by the five-step approach that Minister Li Ganjie came up with, "inspection, assignment, follow-up inspection, admonitory talks and targeted inspection." A complete inspection system has taken shape, whereby central inspectors spot a problem, assign the local government to address it, then follow-up inspections will be conducted to see to the resolution of the problem. We will carry out admonitory talks with local governments that are slow in rectification, and conduct a targeted inspection of those that have major problems but take no rectification action. In the past, we would not know for sure whether a problem spotted is ultimately solved. Now, we need to ensure that all the problems revealed are ultimately addressed, step by step. This is the long-term mechanism we are exploring.

As mentioned earlier, one of the six supporting documents of our stepped-up action for autumn and winter is about quantified accountability. This is an approach proven useful by our explorations on how to ensure that responsibilities are fulfilled all the way down to the community level. If local governments could not complete the tasks assigned to them, we hold them accountable; if we spot problems that previous inspectors missed out on but local governments have not resolved, we hold them accountable. These are good practices extracted from our supervisions and inspections. The mechanisms that prove useful, including the five-step approach I just mentioned, will be replicated in many of the future special law enforcement campaigns. This is what I mean by long-term mechanisms.

Southern Metropolis Daily: I have several questions. First, the stepped-up action plan requires that the "2+26" cities reduce both their $PM_{2.5}$ concentration and heavy pollution days by 15%. How are these targets set, are they related to the assessment targets of the *Action Plan on Air*? Second, how is the reinforced inspection in the stepped-up action plan different from previous reinforced inspection? Also, the stepped-up action plan mentions on-site follow-up inspection. What is the difference between inspection and follow-up inspection?

Liu Bingjiang: Thank you for your questions. You actually touch upon two issues, one is the assessment target of the *Action Plan on Air* and the other the basic principles for setting the targets of the stepped-up action for autumn and winter. It is fair to say that all the targets set in the *Action Plan on Air* had been basically met by the end of 2016, except the one for Beijing, which is reducing its $PM_{2.5}$ concentration to 60 $\mu g/m^3$. Why did we set the targets that the "2+26" cities reduce both their $PM_{2.5}$ concentration and heavy pollution days by 15%? Because people worry and complain the most about the high $PM_{2.5}$ concentration, its long duration and the frequent occurrence of heavy pollution. If these two targets can be met, which means a drop in both overall $PM_{2.5}$ concentration and heavy pollution days, surely we would exceed the targets set in the *Action Plan on Air*.

In setting the targets of the stepped-up action, we followed several basic principles. First, we factor in feasibility. We discussed with localities about this. We broke the targets of the stepped-up action down into quantified tasks for each city. For example, one target is to shut down or suspend 44,000 coal-fired boilers, and each city's contribution to the target is different. It is the same case with the rectification of clusters of enterprises with "location, certificate and emission" issues. Most importantly, the emergency response measures also differ by city. In previous cases of heavy pollution, some cities waited for the wind to blow it away, and some cities only proposed ambiguous measures. But this year, the case is different. Each of the "2+26" cities is tasked with a specific target. For example, if a city is required to cut its nitrogen dioxide emission by 30%, it must do that. We have abandoned the "one-size-fits-all" approach for enterprises and base targets on feasibility.

Second, we factor in local status quo. In the Beijing–Tianjin–Hebei region, cities that emit more pollute other cities. For example, the $PM_{2.5}$ concentration in Shijiazhuang is over 100 $\mu g/m^3$ and that in Jining is over 60. Cities like Shijiazhuang should cut more for the sake of fairness. In a word, the basic principles for setting the targets are feasibility and greater reduction for cities with higher existing concentration. Besides targets set by the central government, local circumstances should be factored in.

Third, we factor in the meteorological transmission patterns. The target for Beijing is reducing its $PM_{2.5}$ concentration to 60 $\mu g/m^3$. However, every time heavy pollution occurs, Beijing's pollution would be aggravated by the wind that blows from southern areas with higher pollution concentration. In the case of smog, it can be blown north by wind from the south or otherwise. Thus transmission patterns also influence pollution concentration. We need scientific measurement and calculations to get the localities involved on the same page about that influence on them, respectively.

Tian Weiyong: Let me answer your second question on the difference between the stepped-up inspection for the "2+26" cities and that for autumn and winter. To put it simply, the latter is clearer in its direction, more specific in its tasks and more precise in the dispatch of personnel. But still, we need to further our efforts for the latter.

Now, let me elaborate. First, a clearer direction. If you pay attention, you would notice that the stepped-up action for the "2+26" cities focuses on industrial enterprises, but the action for autumn and winter has three more specific goals. The first goal is enabling 97% of industrial enterprises to persistently meet the target, which is the bottom line we must stick to. The second goal, also the main one, is cracking down on enterprises with "location, certificate and emission" issues by the end of September and small-scale "scattered" coal burning. Addressing the small-scale "scattered" coal burning involves removing coal-fired boilers and replacing coal with clean energy such as electricity and natural gas. This particularly merits our attention because more coal is consumed in winter. The third goal is resorting to off-peak production and responses to heavy pollution weather as safeguard measures. Prior to heavy pollution, we would step in, issue prewarning and adjust supervisors and inspectors' tasks and carry out a series of supervisions and inspections as our emergency response to heavy pollution.

Second, more specific tasks. The previous stepped-up action mainly has seven aspects including enabling industrial enterprises to gradually meet emission targets and dealing with enterprises with "location, certificate and emission" issues. This time, the stepped-up action for autumn and winter is more specific as it involves 12 aspects, which are subdivided into 233 tasks. Each city is given a specific list of tasks. Inspectors will check them off one by one, and those that are not performed will be considered in future inspections, a process ensuring that all the tasks will ultimately be completed.

Third, more precise personnel dispatch. First, we make daily arrangement for the 28 groups on where to go and what to investigate, and they give feedback every day. Second, we deploy the bidirectional feedback-based supervision and inspection. We have adopted some high-tech means such as the "hotspot grid" and "micro monitoring stations". The squares in a hotspot grid, for example, can be as small as 500 m × 500 m, and they are ranked based on their pollutant concentration. If a square's pollutant concentration is 10 $\mu g/m^3$ higher than others in its periphery, we can be sure that it has a pollution source, which can be a construction site, an enterprise that emits secretly or a problem caused by enterprises with "location, certificate and emission" issues and fail to meet emissions standards. Once the pollution source is identified, we would resolve it. If, however, the pollutant concentration does not come down, it means we are on the wrong track. Then we will continue to investigate and pin down another possible problem. This process will be repeated until we find the real problem, address it and bring down the pollutant concentration. This bidirectional feedback-based approach will enable us to address the problems in a more accurate and targeted way in the Beijing–Tianjin–Hebei region.

"To further our efforts" means making law enforcement more robust. Previously law enforcement has already been perceived to be robust, but it is going to be more robust this autumn and winter. For that purpose, more personnel will be transferred and dispatched. As you know, 5,600 persons have been transferred to carry out inspection tasks. We are in the process of transferring a batch of monitoring staff to carry out spot checks of big enterprises to glean information about their persistent compliance with emission regulations and investigate and penalize enterprises emitting excessively. On top of that, we will assemble special working groups for targeted inspections of key areas, send additional special working groups to areas with more severe problems and, in cases of heavy pollution in winter, dispatch one or several special working groups, depending on the pollution transmission channel. These efforts will become even more intense and you will feel that as well.

Inspection and follow-up inspection have three differences. First, their purposes are different. The five-step approach that I just mentioned, which is "inspection, assignment, follow-up inspection, admonitory talks and targeted inspection," starts with inspection. The purpose of the inspection is to spot problems, which is very important as it is the foundation for the following steps. The purpose of the follow-up inspection, an important part of the five-step approach, is to confirm whether localities have fulfilled the rectification tasks assigned to them and urge them to solve their problems.

Second, their work patterns are different. To put it simply, inspection mostly targets at enterprises. Since April 7, 42,000 enterprises have been inspected, over 20,000 were revealed to be problematic. Follow-up inspection targets at both enterprises and governments. After inspections expose problems in enterprises, inspectors handover the responsibility to local governments. Then it is the task of follow-up inspection to see if enterprises have rectified those problems and, by extension, if local governments have fulfilled the responsibilities handed over to them. In a nutshell, inspections, in the manner of law enforcement, seek to find out if enterprises infringe upon laws and regulations; follow-up inspections aim at finding out if enterprises have rectified their problems and how local governments behave in the process. To find out how local governments fulfill the tasks, follow-up inspectors attend meetings of party committees and standing committees and government executive meetings. You can see that their work patterns are completely different.

Third, their personnel composition is different. Inspection personnel is transferred from localities nationwide, including the 5,600 environmental law enforcement persons. The follow-up inspection personnel are staff from MEP or its affiliated agencies. For this follow-up inspection, we will deploy 1,240 persons and each of them will engage in two rounds. That means altogether 2,480 key on-site follow-up inspectors. Plus, they will be sent to and stay at the "2+26" (prefecture level) cities and counties (county-level cities or districts) under them. We refer to such follow-up inspection as on-site follow-up inspection.

People's Daily: Some people argue that the wet flue gas desulfurization aggravates heavy pollution. What do you think?

Liu Bingjiang: I am aware of that argument and I would respond with several observations. In China, the wet desulfurization approach has always been overshadowed by doubts. Only this time the doubt is shared by more people and spreads further. We are aware of this and I would like to clarify the matter. First, at present, wet desulfurization is the main technology for desulfurization in coal-fired power plants. It is the common choice made by environmental engineers and scientists from all over the world after more than 40 years of R&D, application and upgrading. The share of wet desulfurization application is 98% in Japan, 92% in the United States, 90% in Germany, and 91% in China, averaging 85% in the world. Wet and dry desulfurization is applied under different circumstances. Large coal-fired units use wet desulfurization, while smaller ones with a capacity of less than 50,000 kW use the dry method because the dry method is enough for them. When the doubt about wet desulfurization surfaced, we consulted with our German and American counterparts and they told us that the doubt was groundless.

Second, views need to be corroborated by data. Indeed there are always doubts and questions on desulfurization in China. In 2012, questions on the emission of sulfur trioxide and wet flue gas during the desulfurization process in coal-fired power plants were published on various journals and magazines. Since that year, we have been testing and measuring the pollutants and followed up on every problem. A 60% removal efficiency of sulfur trioxide is what the natural ultra-low emission target requires, and the Beijing–Tianjin–Hebei region and East China have basically reached that percentage. Now the concentration of sulfur trioxide does not exceed 10 mg/m^3 and averages several milligrams per cubic meter, which is equivalent to one-tenth of sulfur dioxide and hence negligible. Regarding soluble salts, MEP estimates that the weight of such salts stands at 10,000 tons, which, compared with the 10 million tons of sulfur dioxide emitted nationwide every year, is completely negligible. As for the wet flue gas problem, I can tell you with certainty that the flaws found in the wet desulfurization process also exist in the dry desulfurization process, just to a different degree. With wet desulfurization, the flue gas emitted is at over 50 °C. The flue gas from dry desulfurization, if emitted at over 50 °C, will look like white smoke billowing up into the sky, just like that from wet desulfurization. The water vapor from wet desulfurization is 10% more than that from dry desulfurization. Some people argue that to rein in haze, the flue gas should be heated before emission. But I can tell you with certainty a critical message that the absolute amount of water vapor is the same whether you heat it or not, though when you heat it, it becomes visible to naked eyes. Scientists' calculations show clearly that the amount of wet flue gas pales in comparison with that of the water vapor in the nature, thus its influence on haze is negligible.

Third, some environmentally friendly Chinese power companies are very angry that dry sulfurization is believed to trump wet desulfurization not based on scientific evidence but some data that is pieced together. The Chinese engineers dedicated to environmental protection in power have developed the most advanced ultra-low emission approach in the world, and the concentration of sulfur dioxide in China's coal-fired plants has dropped to 20–30 mg/m^3, the lowest in the world. They are proud of the achievement, but do not feel they have been recognized in a way they deserve. What I just said might give them some comfort. China Electricity Council is hosting a meeting on the 19 this month, to which major environmental protection engineers in state-owned companies and their German or U.S. counterparts will be invited. People of different views are welcome to talk to each other.

Southern Weekly: One of the six supporting documents for the stepped-up action plan is on publicity. So what will the arrangements for publicity be?

Liu Youbin: MEP attaches great importance to the publicity of the stepped-up action, and thus incorporates it into the overarching plan for the stepped-up action and makes thorough and holistic arrangements on it. MEP will, via press conferences, briefings, press releases and interviews by accompanying media, vigorously promote the importance, necessity, justification, and urgency of the stepped-up action so as to build consensus among the general public and create favorable public opinion for improving the atmospheric environment quality. We will organize the timely coverage of models. In cases of insufficient progress of local governments in the comprehensive treatment of air pollution and illegal behaviors that severely infringe upon the public's environmental rights and interests, the media as watchdog will play a stronger role. At the same time, we will vigorously disseminate laws, regulations and scientific knowledge on environmental protection, and mobilize the public to engage in environmental protection and reject illegalities. MEP's official WeChat public account publicizes the progress of the stepped-up action every day.

We ask localities to highly value public opinion, timely release authoritative information, actively respond to the public's concerns and innovate in communication methods. In this way, the central and local governments can join hands and synchronize their efforts, cooperating with the media for their interviews and coverage of the stepped-up action.

Supportive public opinion is an important safeguard to a smooth stepped-up action for comprehensively treating air pollution in autumn and winter, which cannot do without the vigorous support of the media friends. We will facilitate and support the coverage and interviews of media professionals. Together, we will create a new normal for environmental law compliance and protect the blue sky.

Liu Youbin: That is all for today's press conference. Thank you!

Press Conference on the 13th Five-Year Plan for Nuclear Safety and Radioactive Pollution Prevention and Control and the 2025 Long-Term Targets

March 23, 2017

Recently, the State Council has given an official reply to the *13th Five-Year Plan for Nuclear Safety and Radioactive Pollution Prevention and Control and the 2025 Long-Term Targets* (hereinafter referred to as "the plan"). Personnel in charge from MEP (NNSA) answered journalists' questions on it.

Ministry of Environmental Protection of the People's Republic of China,
2017 Press Conference Records of Ministry of Environmental Protection, the People's Republic of China, https://doi.org/10.1007/978-981-13-7330-5_15

Press conference (1)

Press conference (2)

Question: Why was the plan formulated? What is the significance of launching it?

Answer: Nuclear safety, as well as radioactive pollution prevention and control, have a bearing on people's health, environmental safety, and social stability. The CPC Central Committee and the State Council attach great importance to nuclear safety and radioactive pollution prevention and control, integrate nuclear safety into the country's overall security system and elevate it as a national security strategy. At the Nuclear Security Summit, General Secretary Xi Jinping proposed the "rational, coordinated and balanced" concept of nuclear safety and solemnly pledged to the world that China would formulate a mid- to long-term nuclear safety development plan. The *Outline of the 13th Five-Year Plan for National Economic and Social Development* (hereinafter referred to as "the outline") specified improving nuclear facilities safety, preventing and controlling radioactive pollution, and strengthening the regulatory system and capacity building for nuclear and radiation safety. Many other central documents have also proposed formulating a nuclear safety plan. General Secretary Xi Jinping and Premier Li Keqiang, among other central leaders, have also issued important instructions on nuclear safety on many occasions, stressing the need to do everything possible to eliminate hidden dangers.

The Ministry of Environmental Protection (MEP) (the National Nuclear Safety Administration, NNSA) took the lead and together with the National Development and Reform Commission, the Ministry of Finance, the National Energy Administration and the State Administration of Science, Technology, and Industry for National Defense, completed the plan after meticulous research, extensive solicitation of opinions, and scientific and specific debates over three years. The purposes are to fulfill the arrangements and requirements of the central government, coordinate nuclear safety as well as radioactive pollution prevention and control during the 13th Five-Year Plan period, and ensure the safe and efficient development of China's nuclear energy and nuclear technology utilization. The plan fully reflects the weight given by the CPC Central Committee and the State Council to nuclear safety and the prevention and control of radioactive pollution, President Xi Jinping's nuclear safety concept, which is "rational, coordinated and balanced", the fundamental principle of equally valuing safety and development, and the weight given by the CPC Central Committee with President Xi Jinping at the core to people's vital interests. The compilation and implementation of the plan are of great significance to enhance management, improve the safety of nuclear facilities, reduce risks, promote the prevention and control of radioactive pollution, maintain sound radiation environment quality, shore up people's confidence, promote nuclear power overseas, and implement the Belt and Road initiative.

Question: What is the plan like? Anything innovative about it?

Answer: The plan is an important part of national security top-level design and of the strategic arrangements for ecological and environmental protection. It is dedicated to guiding and strengthening China's efforts on nuclear safety and radioactive pollution prevention and control. It serves as a safeguard for the safe and healthy development of nuclear energy and nuclear technology utilization. Drawing upon experience, the plan makes breakthroughs and innovations and holistically coordinates nuclear safety and radioactive pollution prevention and control. The plan mainly has the following characteristics:

First, the nuclear safety concept is the overarching principle. The plan equally values development and safety, emphasizes defence in depth and continuously takes improvement measures for nuclear safety. The plan pays equal attention to rights and obligations. It delineates responsibilities for all parties concerned including central authorities, local governments and companies, and public institutions, which speaks volumes about social co-governance. The plan addresses both symptoms and root causes. The operation of nuclear facilities and decommissioning of old ones are both covered. Independence and cooperation are reconcilable as the plan stresses learning from international experience and formulating targeted measures based on China's reality at the same time.

Second, the protection of health and eco-environmental safety is the fundamental purpose. The plan meets requirements of the CPC Central Committee and the State Council regarding eliminating hidden dangers that affect nuclear safety and protecting health and eco-environmental safety. The fundamental guiding principle is "safety first, quality first", and the purpose is to "maintain sound radiation environmental quality and effectively and constantly safeguard nuclear safety, environmental safety and public health." People's health and eco-environmental safety constitute as the fundamental starting point of our work during the 13th Five-Year Plan period.

Third, new development concepts are the guide. Committed to innovation and reform as the driving force, the plan puts forward a series of tasks and measures such as speeding up management innovation, institutional reform, and technological innovation. Committed to coordinated development, the plan proposes continuing with improvement measures to make China one of the leading countries in the overall safety of nuclear facilities. Committed to green development, the plan proposes accelerating the decommissioning of old nuclear facilities and radioactive waste management to reduce environmental risks. Committed to openness, the plan proposes drawing upon international experience and lessons, sharing China's good practices and making international exchanges and cooperation in this regard more extensive and in depth. Committed to the sharing of benefits, the plan pays more attention to building an open and transparent nuclear safety regulatory system and strengthening science popularization and publicity, information disclosure and public participation.

Fourth, risk prevention and control is the core. Geared toward solving problems and mitigating risks, the plan painstakingly analyses the major risks that may face China's nuclear energy and nuclear technology utilization. It sets targets, assigns key tasks, establishes key projects, and puts forward safeguard measures, all for the purpose of risk mitigation.

Fifth, capacity building is the support. The plan focuses on capacity enhancement in scientific research, emergency preparedness and nuclear safety regulation at the central, provincial, prefecture, and municipal levels. Regarding scientific research, the focus is on tackling key problems affecting nuclear safety; regarding emergency preparedness, we will consider both soft and hard skills and combine command and technology in peacetime and wartime alike. Regarding regulation, the focus is on such general aspects as evaluation and approval, law enforcement, radiation monitoring, experience feedback, communication with the public and international cooperation.

Sixth, improvement in nuclear and radiation safety is the goal. Goals set in the plan include the following ones: the safety level of China's nuclear facilities both operating and under construction would be significantly improved during the 13th Five-Year Plan period; the safety of nuclear power would be among the best in the world; the incidence of radiation accidents from radioactive sources would be further reduced; no nuclear accidents resulting in radioactive pollution of the environment would occur; by 2025, the safety of China's nuclear power plants would maintain an internationally leading edge, and that of other nuclear facilities will reach an internationally advanced level, and the incidence of radiation accidents involving radioactive sources will remain at a relatively low level.

Question: Could you please elaborate on the plan? What are its new requirements and measures?

Answer: The plan can be summed up as six goals, ten key tasks, six key projects, and eight safeguard measures. Specifically speaking, the six goals are the improvement in six aspects, namely safety of nuclear facilities, safety of nuclear technology utilization devices, radioactive pollution prevention and control, nuclear security, nuclear and radiation emergency preparedness and safety regulation.

The ten key tasks are maintaining a high level of safety in nuclear power plants, reducing the risks to research reactors and nuclear fuel cycle facilities, speeding up the decommissioning of old nuclear facilities and the disposal of radioactive waste, reducing the incidence of radiation accidents caused by the use of nuclear technology, and ensuring the safety of uranium mining and milling and the radiation environment of radioactive associated minerals, improving the quality and reliability of nuclear safety equipment, enhancing nuclear security, strengthening nuclear and radiation accident emergency preparedness, promoting R&D efforts on nuclear safety and modernizing nuclear safety regulation. The six key projects involve nuclear safety improvement, nuclear facility decommissioning and radioactive waste disposal, nuclear security, and anti-terrorism upgrade, emergency preparedness for and safeguards against nuclear accidents, scientific and technological innovation in nuclear safety, and nuclear safety regulatory capacity building. The eight safeguard measures are perfecting laws and regulations, strengthening supporting policies, optimizing system and mechanism, speeding up talent cultivation, fostering greater awareness, promoting communication with the public, deepening international cooperation, and improving investment mechanism.

The specifics of the plan, be it goals, key tasks, key projects, or safeguards, all fully manifest the guiding thought of developing nuclear energy and technology on the basis of ensuring safety and the fundamental principle of "safety first, quality first". The plan has also come up with a series of new requirements: developing nuclear energy and utilizing nuclear technologies in accordance with the latest international safety standards, reinforcing the concept of managing nuclear according to the law, and enforcing the strictest nuclear safety regulation. Nuclear energy development departments, safety regulatory authorities, people's governments at all levels, enterprises, and public institutions should earnestly fulfill the fundamental purpose of safeguarding nuclear safety, environmental safety and public health. The plan has also put forward new measures on strengthening the rule of law, institutional and mechanism building, team building, safeguard capacity building, and a nuclear safety culture.

Question: What achievements were made in nuclear safety and radioactive pollution prevention and control in China during the 12th Five-Year Plan period?

Answer: In 2012, the State Council has deliberated and approved the implementation of the *12th Five-Year Plan for Nuclear Safety and Radioactive Pollution Prevention and Control and the 2020 Long-Term Targets*. Thanks to the joint efforts of competent central authorities, local governments, and the entire industry, positive results were secured during this period.

Press conference (3)

First, nuclear facilities safety has been improved. Nuclear power safety has reached an internationally advanced level. The safety performance indicators of operating nuclear power units, compared with those of the same kind worldwide, rank among the top; the quality of units under construction is within control, and the design indicators of the newly built nuclear power units meet the latest international nuclear safety standards; sound measures to prevent and mitigate serious accidents have been put in place, and potential dangers facing research reactors and nuclear fuel cycle facilities have been eliminated.

Second, phased progress has been made in the prevention and control of radioactive pollution. The decommissioning of some old nuclear facilities has been completed. China's capability in processing and disposing radioactive waste has been improved, and northwestern, southwestern and southern China have developed their independent disposal capacity. Some legacy radioactive waste has been disposed. We have completed the decommissioning of a number of uranium mining and milling facilities, and basically completed the decommissioning and treatment of uranium geological exploration facilities in key areas. Thus, environmental risks have been continuously reduced.

Third, the incidence of radiation accidents caused by radioactive sources has continued to decrease. The annual incidence of radiation accident from radioactive sources has dropped from an average of 2.5 per 10,000 radioactive sources during the 11th Five-Year Plan period to less than 2 now, the lowest level in history. There have been no particularly serious radiation accidents, and all kinds of used radioactive sources have been collected and stored in a timely manner, thus ensuring the safety of people and the environment.

Fourth, the nuclear safety safeguard system is becoming increasingly sound. The legislative process of *Nuclear Safety Law* is speeding up; nuclear safety administrative agencies are expanding and staffed with more people. We have started construction of the national nuclear and radiation safety regulatory technology R&D base and have basically put in place a national environmental radiological monitoring network. We have built 21 major scientific and technological innovation platforms, and carried out and made key breakthroughs in 200-odd studies on nuclear safety technologies. We have improved the emergency preparedness system, carried out nuclear emergency preparedness capacity building and established an emergency preparedness force under unified control.

Question: Could you please brief us about the current nuclear energy and technology utilization scale and the overall situation of nuclear safety in China?

Answer: China is a big utilizer of nuclear energy and technology. Currently, it has 35 nuclear power units operating and 21 under construction, the latter ranking first in the world. It also has 19 research reactors and nearly 100 nuclear fuel cycle facilities. Altogether 67,000 institutions in China utilize nuclear technologies with 127,000 radioactive sources and 151,000 radiation devices (sets) being in use, and 192,000 disused radioactive sources have been recovered.

China has safely utilized nuclear energy and technologies for over three decades without any accidents rated at or above Level 2 on the International Nuclear Event Scale. China's nuclear safety is world leading. The incidence of radiation accidents caused by a radioactive source is dropping, and research reactors and nuclear fuel cycle facilities keep good safety records. Overall, the risks posed by nuclear and radiation to safety are controllable, the environment radiation nationwide is within the natural background radiation range, and no radioactive pollution events have occurred. China has developed a comprehensive supporting capacity for accident prevention, pollution control, technological innovation, emergency preparedness, and safety regulation, effectively safeguarding nuclear safety, environmental safety, and public health.

Question: What are the plan's guiding principles for China's nuclear safety and radioactive pollution prevention and control during the 13th Five-Year Plan period?

Answer: According to the content of the outline and the instructions of the CPC Central Committee and the State Council on nuclear safety and based on the actual situation in China, the plan determines the guiding principles of nuclear safety and radioactive pollution prevention and control during the 13th Five-Year Plan period. We should implement the guidelines from the 18th CPC National Congress and its 3rd, 4th, 5th, and 6th plenary session, follow the guidance of Deng Xiaoping Theory, the Important Thought of Three Represents and the Scientific Outlook on Development, put into practice the principles conveyed in the series of important speeches delivered by General Secretary Xi Jinping and the New Ideas, New Thoughts and New Strategies on Governance. We should implement the decisions and plans made by the CPC Central Committee and the State Council, and coordinate efforts to pursue the "five-in-one plan"[1] and the "four-pronged comprehensive strategy".[2] We should promote innovative, coordinated, green, open, and shared development, and advocate a rational, coordinated and balanced nuclear safety outlook. We should adhere to the "safety first, quality first" principle, and view risk prevention as the core, law-based nuclear governance as the foundation, nuclear safety culture as the guidance, reform, and innovation as the driving force, and capacity building as the support. Actors and entities concerned should fulfill their responsibilities, continuously improve safety and enhance radioactive pollution prevention and control to safely and efficiently utilize China's nuclear energy and technologies.

Question: What are the plan's goals for China's nuclear safety and radioactive pollution prevention and control during the 13th Five-Year Plan period?

Answer: China's goals for 2020 are set explicitly, which, generally speaking, are as follows: safety of nuclear facilities both operating and under construction will be significantly improved; nuclear power safety will remain world-leading; radiation accidents caused by radioactive sources will be reduced; decommissioning of old nuclear facilities and radioactive pollution control will be fruitful; no radioactive pollution accidents will occur; the radiation environment maintains good quality; nuclear emergency preparedness and safety regulation will be improved; nuclear safety, environmental safety, and public health will be safeguarded. Specifically speaking, the goals are an improvement in six aspects: nuclear facility safety, nuclear technology utilization device safety, radioactive pollution prevention and control, nuclear security, nuclear and radiation emergency preparedness, and safety regulation.

Question: How is China's nuclear and radiation safety regulation? What are the plan's measures for improving regulation?

Answer: China always attaches importance to nuclear and radiation safety regulation. The National Nuclear Safety Administration (NNSA), a regulatory agency independent of the nuclear energy development department, is established to regulate site selection, design, construction, operation and decommissioning of

[1]Coordinated progress in economic, political, cultural, social, and ecological areas.

[2]Building a moderately prosperous society in all respects, deepening reform, fully implementing the rule of law, and strengthening party discipline.

nuclear facilities. China's nuclear safety regulation is consistent with international practices.

Over the past three decades, China has optimized its nuclear safety regulation system and improved regulation. China has put in place a system and a team with expertise to regulate nuclear safety according to the established laws. Since the 12th Five-Year Plan period, China has made breakthroughs in its nuclear and radiation safety regulation capacity: the *Nuclear Safety Law* was approved in the first round by the Standing Committee of the National People's Congress; a national regulation technology R&D base started construction; a national environmental radiological monitoring network has basically taken shape; on-site supervision and law enforcement equipment has been improving, and local regulation capacity has been strengthened. In August and September 2016, the International Atomic Energy Agency (IAEA) made a comprehensive follow-up assessment of China's nuclear and radiation safety regulation and concluded that China safeguarded public health and environmental safety by publishing and implementing nuclear safety plans and giving more human and financial resources to nuclear and radiation safety regulators. China's regulation is in line with international practice, and the regulatory agencies are effective and reliable. During the 12th Five-Year Plan period, they improved legislation and strengthened internal management, independently and effectively carried out regulatory activities, established supervision and monitoring systems and swiftly and effectively acted upon lessons learned from the Fukushima nuclear accident in Japan. These distinctive measures are worth promoting globally.

The third plenary session of the 18th CPC Central Committee proposed modernizing the national governance system and capacity, pointing out the direction for the nuclear safety regulatory system and capacity building. Regarding tangible facilities, infrastructure and the like, the plan proposed three measures. First, an R&D base for regulation technologies will be built to improve our independent calculation, experimentation, and verification. Second, apparatuses and equipment in regional supervision stations and provincial regulation agencies will be improved to boost on-site supervision and law enforcement. Third, the national environmental radiological monitoring network will be improved. To that end, monitoring capacity building will be carried out at the central, provincial, and key-prefecture-level, respectively, with the emphasis placed on key ports and border areas. Also, along with nuclear safety regulation, nuclear power and technology utilization will be advanced to modernize the country's safety regulatory system and capacity by 2025.

Question: What scale will China's nuclear power reach in the 13th Five-Year Plan period? Some experts and media outlets argued that treating spent fuel and radioactive waste would be a major challenge. What specific measures does the plan have in response?

Answer: According to the medium- and long-term plan for nuclear power, by the end of the 13th Five-Year Plan period, China's total nuclear power units will rank second in the world, with the installed capacity of operating units being 58 million kilowatts and that of units under construction being over 30 million kilowatts. Indeed treating spent fuel and radioactive waste from nuclear power plants has a major bearing on China's nuclear power development.

China always values spent fuel safety and has established a closed-loop recycling policy for spent fuel reprocessing to make full use of fissile materials. To that end, China set up a fund dedicated to spent fuel reprocessing to ensure it would not be handicapped by funding shortage. Also, China has vigorously promoted and made breakthroughs in the R&D of reprocessing technologies. The pilot plant for reprocessing power reactor spent fuel is under hot test. China is negotiating with France to jointly build a large-scale commercial reprocessing plant, and China National Nuclear Corporation is conducting the first round of site screening. The reprocessing plant must meet the strict technical requirements of and be approved by MEP (NNSA) before construction. Safety and environmental protection requirements imposed on reprocessing plants are on par with those on nuclear power plants. Overall, China's spent fuel storage is safe and stable. However, the reactor storage capacity of some nuclear power plants comes short and their spent fuel is in dire need of shipping out. In light of this, as stated in the plan, during the 13th Five-Year Plan period we will formulate and release the plan for treating spent fuel to solve storage and reprocessing problems. We will, according to the law, specify safety requirements that near-reactor dry storage facilities must meet to be approved and speed up away-from-reactor storage capacity building. We will strengthen the top-level design for stronger industry–university–research cooperation, establish a safeguard mechanism and optimize operation and management, vigorously push forward the site selection and construction of large commercial spent fuel reprocessing plants to relieve at-reactor storage pressure.

China also high values of radioactive waste treatment. Thanks to industry-wide engagement and efforts, the 12th Five-Year Plan period saw great progress in radioactive waste management. However, some radioactive waste remains untreated because of multiple reasons, the main one being that treatment capacity comes short and cannot keep up with nuclear industrial development pace. Therefore, as stated in the plan, we will accelerate reactive waste treatment capacity building, basically complete the solidification of the remaining low- and intermediate-level radioactive liquid waste, and treat part of the low- and intermediate-level radioactive solid waste. We will publish and implement the *Plan for Disposal Sites of Low- and Intermediate-Level Radioactive Solid Waste*, select locations for and construct five such disposal sites so that a proper layout can take shape, and facilitate the shipping out and disposal of nuclear waste. We will start constructing underground research laboratories for geological disposal of high-level radioactive waste, facilitate location selection and survey, and accelerate research on high-level radioactive waste disposal.

Question: What are the plan's measures to promote communication with the public about nuclear safety to safeguard their right to be informed, participate, and supervise?

Answer: MEP (NNSA), as China's independent nuclear safety regulator, values the building of an open and transparent regulation system. During the 12th Five-Year Plan period, NNSA vigorously promoted the publication of environmental assessment reports, national environmental radiological monitoring results, project approval results and relevant documents about nuclear projects. It also disseminated nuclear safety knowledge on campuses, in communities and training classes for officials to boost public engagement in the project construction phase. Communication with the public has been fruitful.

As stated in the plan, during the 13th Five-Year Plan period, we will promote communication with the public about nuclear safety based on the tenet of "central supervision and guidance, local leadership, corporate action and public participation." We will stick to openness and transparency as one of the basic principles, fulfill our responsibilities of communicating with the public, improve the communication mechanism and safeguard the public's right to be informed and participate according to the law. To that end, during the 13th Five-Year Plan period, we will incorporate basic nuclear safety knowledge into education and training systems and disseminate nuclear and radiation knowledge in communities, primary and secondary schools and training classes for officials. We will rely on enterprises to build ten national nuclear and radiation safety education bases, bring the Internet and new media into better play and strengthen our exchange with the media. We will improve information publicizing plans and guidelines and establish platforms for publicizing information. Enterprises shall disclose project progress information phase by phase according to the law, and the government shall actively disclose information on project approvals, supervision and law enforcement, environment monitoring results and accidents and events, and offer more interpretation on disclosed information.

MIX
Papier aus verantwortungsvollen Quellen
Paper from responsible sources
FSC® C105338

If you have any concerns about our products,
you can contact us on
ProductSafety@springernature.com

In case Publisher is established outside the EU,
the EU authorized representative is:
Springer Nature Customer Service Center GmbH
Europaplatz 3, 69115 Heidelberg, Germany

Printed by Libri Plureos GmbH
in Hamburg, Germany